OVER THE DISTANT HILL

OVER THE DISTANT HILL

Robert Lawson

The Book Guild Ltd
Sussex, England

First published in Great Britain in 2005 by
The Book Guild Ltd
25 High Street
Lewes, East Sussex
BN7 2LU

Typesetting in Times by
Acorn Bookwork Ltd, Salisbury, Wiltshire

Printed in Great Britain by
CPI Bath

A catalogue record for this book is available from
The British Library.

ISBN 1 85776 899 X

CONTENTS

ACKNOWLEDGEMENTS

When compiling a record of events that occurred many years ago the writer must rely almost entirely upon the quality of the recollections and anecdotes sent for inclusion. Fortunately less than half of the reminiscences were from people with sharp, analytical minds, so I never felt out of my depth. Sometimes the same object or event was recalled a little differently, like the grocer's delivery vehicle, which could have been a 1936 Citroën, a nearly new Morris 1000 van, or a Wolseley Wasp. Similarly, when people were asked how many black huts there had been, the answers varied from eight to twenty. Well, at least they remembered there were huts.

My thanks to all who took the time and care to delve back into their archives and managed to avoid the wardrobe full of rattling skeletons so adroitly. Names are in the order that recollections were received. So my appreciation to:

Anita, for being the quickest to respond by about two weeks.

Ted, for sending most photos and remembering more than anyone else.

Queenie, my mum, for creating most of the people who replied.

Tom, for being the source of so many memories.

Pat, for showing me there is always an alternative viewpoint.

Eileen, for reading a chapter without being asked.

Richard, for helping me to see how unimportant some things are.

Barbara, for not putting the answerphone on when I called again.

René, Sarah and Alexandra, my children, for continually listening as if they were hearing it for the first time.

Finally, to my wife Ann, within whose all-embracing vision no pessimistic thought dared linger, my warmest love and gratitude.

INTRODUCTION

About a year ago my cousin Ted telephoned to say that he was tracing back his family and needed some information. We spoke for a long time, mostly about the summers we spent between 1953 and 1960, hop-picking in Kent. We had so many memories between us that my wife, Ann, suggested it would make them easier to recall if I wrote them down.

Over the next few weeks I contacted all of my family for their ideas and recollections. Reactions varied from, 'Oh, really? I'll phone you', to an animated hour-long chat and three pages of memories. I had intended to write just about the days spent on the hop farm, but that seemed so inseparable from the rest of my life, like Latton Lock Cottage, 90 St Ann's Road and Soper's farm, that I have included those events that drew some emotion from me and hope that people reading it will experience the same feelings.

Readers should first ask themselves how they would feel, spending one and a half months each summer in a windowless, wooden shed, sleeping on a straw filled mattress and cooking on an open fire. When almost freezing water of unusual taste is supplied through a standpipe one hundred yards away, shared by thirty-five people. When a hole in the ground inside a Portakabin serves as a community toilet, with no possible chance of a bath or shower. There are no televisions, no computers, no telephones. The day is spent picking hops the size of kidney beans into a huge basket and the sharp bines can cut soft skin like razors.

Does this sound like paradise or purgatory? For children from London in the 1950s it was invigorating. It was fantastic, it was 'Hopping Down in Kent' as the song goes.

For those six weeks we lived in a completely different environment, where freedom and fun appeared high on the agenda of daily duties as opposed to constantly obeying rules. Apart from a few hours working, the day's activities were decided by us alone and opened a door to a myriad of new sights and sounds.

Of all the memories contained here, the one event that everyone without exception recalled was of sitting around the fire on Friday evenings, waiting for Dad's car lights to appear over the distant hill. Thomas Hardy, in *His Immortality*, noted that as long as there is even one person still living who knew you and thinks of you, then part of you lives on in them. Dad, you live on in us all.

1

The Last Day

The sixth of September 1960. It was the sort of evening when pockets are essential, both for keeping fingers free of frostbite and for carrying small trowels. My cousin Ted and I were toasting ourselves by a huge log fire, where everyone was having a sing-song. The youngsters bellowing the latest hits, followed by the adults' version of 'The White Cliffs of Dover'. They had just swung into an Alma Cogan number when I nodded to Ted and we strolled into the darkness, attracting as little attention as possible, though as usual in the early evening, squadrons of bats seemed intent upon testing their radar out on us, skimming at head height, then veering away at the last moment.

We began jogging down towards the gate, each carrying a sack and clearly off scrumping, but this time it was for vegetables. At the side of the farmhouse there was a kitchen garden, and we rapidly filled our sacks with an assortment of potatoes, beans, lettuces and cabbages. There was an excitement about it that made me tingle. We were about to move round to the back of the house, when Ted pointed up the road to the distant hill. A tiny light had appeared and was approaching the fork in the road, where it would either go left for Canterbury, or right for Petham and the black huts. It was a motorbike and the only person who rode one here was my older brother Tom, who should be in Harlow. A shiver ran through the hairs on my neck as the light took the right fork, and I picked up my half-full sack and ran for the gate with Ted close behind.

As we reached the road the bike was just entering the field. It was a red and cream Matchless G12 De Luxe being ridden by Tom. He was nineteen, six feet tall and stocky, even at that age. He took fair play to an almost religious level. One often repeated tale concerns a time early in his Sunday football career when he was captain of a well-known local factory team, and one of his players made such an awful tackle that, as the referee appeared not to have seen it, Tom decided he should do the ref's job and sent the player off. He has a passion for motorbikes and was an original 'Ton-up kid'.

Whenever I rode pillion with him I felt as if I was in a centrifuge – one of those fairground type of rides that is used in training programmes for pilots and astronauts. It activates the face, the G force making the corners of the mouth touch the ears. Yet I still had total confidence in his riding ability.

The youngsters from the camp were running towards him, excited by this unexpected event. By the time Tom pulled in beside the huts and lifted his bike up on to its footrest, I was halfway across the field, still clutching the heavy sack, but now I almost slowed to a stop. My neck prickled again with anticipation of bad news. As Tom reached the rest of my family, who were grouped outside the hut, he obviously said something devastating, because everyone stopped moving for a few seconds, as if unsure of what to do next. Then Barbara, the youngest, screamed and ran off. Fortunately she ran towards me, and the shock of me appearing out of the darkness made her stop crying, turn around and run back to the hut. Someone had brought my mum a chair and she fell back into it.

I guessed what had happened. My dad, who was only forty-four, had been ill for a little while and although most of us didn't know what Hodgkin's disease was, we were aware that he was not getting any better. He had insisted on us all going to Petham that year and he was taken into St Margaret's Hospital in Epping, Essex. Pat, my oldest brother, then aged twenty-one, Tom and Anita, who was seventeen, visited him. They all had jobs back in Harlow, so they only came down to the hop fields on the occasional weekend.

I walked slowly over to the group and said, 'It's dad, isn't it?'

Tom nodded. 'He died this afternoon.'

Although I had half expected the answer, it still came as a thump in the chest. I can't explain my reaction, but it was to throw the sack I carried into a corner and say, 'Well, we won't be needing these then.'

I have no recollection of how we all got back home. Tom is sure that he stayed the night, then took mum to Harlow, before going on to the hospital in Epping. Aunt Marge looked after the children until the next day, when Tom came down again, driving a 1957 Vauxhall Wyvern that had belonged to dad, though he had not been able to use it recently.

Anyway, after leaving Petham my next recollection was of entering our front door and mum telling one of us to make sure all the curtains were kept closed. To my knowledge none of us ever went to Collard's farm again, unless it was years later in search of their roots.

2

The First Day

It was just the day for seeing new faces and places. The early morning air had painted a pink tip on the noses peering from under the hats that people found essential in 1953. Had there been a name inside the green Robin Hood number, it would have read *Bob Lawson*. It was the last week in August and the Lawson family were, for the first time, going to spend six weeks in the fresh air of Kent, picking hops.

All nine of us had arrived in Edith Road, Leyton, the night before, having spent that day shuttling back and forth from Tottenham, taking small pieces of furniture, crockery, cutlery and linen that would be needed in our temporary home. We spent the night trying to sleep on the floor of Jim and Marge Hylott's house. They were my uncle and aunt, Ted's parents. The house had four rooms, a small yard, and an outside toilet, so when we were called at 6 a.m., most of us were more than ready to get up and stretch our cramped muscles.

As soon as we had dressed, Ted and I walked to the shop on the corner of Edith Road for the *Daily Sketch* and some tobacco for my dad. We were held up as usual by the need to search around the bomb site where three houses in the terrace had been completely wrecked in the war. Most of the kids around there spent a lot of time making dens, lighting fires and looking for anything of value, like metal, woollens or cardboard, that we sold to the rag-and-bone men who were always pushing their carts around the area. Sites like this were common in the fifties, and even if no riches were found there, they were good fun to play on.

The removal van was expected at 7 a.m., so we began transporting what seemed like all of the house contents out onto the pavement, to be picked up by the removal men when they arrived. The removal company we used was called R. Open, from Leytonstone, and they are apparently still operating to this day.

While this was going on, my cousin Ted, who was eight, and a person who liked to be involved, had waited until I had delivered the newspaper and tobacco, then gone to the other corner to give a wave if he saw the lorry coming. I decided to go along to the next junction, to give an even earlier warning. The problem facing me was that I wasn't too sure what to look out for, so I spent the next half an hour imagining I was training for the decathlon and went through all of the track and field events, until I was satisfied I would win gold in the next Olympic Games. Then Ted ran up to me, shouting that they were almost ready to leave. I was peeved, not only to have missed seeing the lorry first, but at not being there in time for the loading, although in the excitement that was soon forgotten.

We were the last group to be picked up and the inside of the van was a maze of furniture, with piles of linen and blankets strewn around for people to sit on. One of the back doors had been strapped back and the adults were all sitting round the opening, apparently singing songs and waving to the overtaking motorists. The children were all spreadeagled about on the blankets. Ted and his nine-year-old sister Pauline were the only ones I knew, but before the next six weeks were through I would look upon most of them as good friends, especially nine-year-old Arthur Lewis, who, along with his brother Ray aged seven, spent most of the time in Kent with me and Ted. At that moment the brothers were, I think, rowing an upturned table across the Atlantic Ocean in a race with Eric Pearce, who was eight, and Eileen, my six-year-old sister. This pairing had the advantage of a seaworthy open-doored wardrobe which happened to be nearer the tail of the lorry, meaning they would be unloaded before the table.

This seagoing saga in a removal van added another dimension to what was already a most unusual journey. Many of the hop-pickers from London must have had a yearning to be performers, because they spent much of their time singing, especially on that long journey to Collard's farm. Barber-shop, soul, even hymns, you name it, they tried to sing it, while waving to anyone who was watching, which probably numbered quite a few as it's not often that you see a removal van being driven along the high street harmonising 'Roll out the Barrels'. Still, if success can be measured by the pleasure attained, that crowd in R. Open's removal truck succeeded all the way to Petham. One or two of the parents did set themselves up as guards, shouting at any child that moved, but mostly they joined in with the others and left us alone.

On this first trip to Petham I only had an occasional glimpse of the towns that we passed through, but the times I've made that journey since, both physically, and mentally by map, have ensured that most of the places are imprinted on my mind.

The Blackwall Tunnel, built by one of the world's most famous engineers, Isambard Kingdom Brunel, has that unique rumble from the traffic that strikes an instant fear into any first-time tunnel traveller. The sound seems to surround the vehicle, like a crumbling wave overpowering a group of surfers, then rebuilding for another plunging attempt to sweep the motorist from the tunnel exit. By the time the end is reached, driving out into the comparatively fresh air and normal sounds is a definite anti-climax. From the tunnel we got on to the A2, and I recall Dartford, where there is now another famous tunnel under the Thames which was opened in 1963, but back in 1953 it was well known for its cement works. Then the Roman town of Rochester with its 12th century castle, and the extensive Chatham dockyards and a naval base that dates from Tudor times. Just before the cathedral city of Canterbury, where Becket was martyred in 1170, there is a turn-off to Chartham Hatch. Then on to the village of Petham. Just outside Petham is Collard's farm, Swarling Manor.

The driver manoeuvred, with some difficulty, through a five-bar gate and across a huge field, pulling up in front of a row of black-painted wooden huts. There were twelve of them, separated in the middle by a brick cookhouse for each end.

The first thing that needed clearing away was the immense amount of nettles that had grown around the huts since they had last been occupied. One of the group remarked that it could almost have been mistaken for Sleeping Beauty's castle. Well yes, except that most of the women there looked like ravaged insomniacs who, rather than having slept for the last one hundred years, hadn't managed forty winks between them, plus the possibility of Prince Charming kissing one of them would lead you to the conclusion that at least four of his five senses were defective.

So many people were helping that the foliage was quickly cleared, and once we'd been allocated our huts, it was a very short time before we had all moved in. Even the cleaning was invigoratingly different.

Meanwhile, a couple of the adults had lit a fire from some of the huge mass of wood that the tractor driver had transported from the spinney. Others, after liberally soaking the earth nearby, used the resultant mud to bake a mass of potatoes and apples. This didn't come too soon for what was quickly turning into a ravenous horde.

It was late afternoon by the time everything had been sorted out, and we barely had time to investigate the local area before dusk arrived, enticing scores of bats from the barn in the corner of the field. There was a row of gypsy caravans on the far side of the field. Some of the black hut group got quite friendly with them, but as our little group was a bit young and there were enough of us not to need any other members, we mostly stayed away from them. We did have a wander around the nearest hop field, because some of us had never even seen a hop before, let alone a fieldful. As we walked between the rows of bines that wound up the strings attached to overhead wires, I think we all felt as if we had stepped into an Enid Blyton story. Especially when we reached the spinney.

The other side of the dense woodland was Petham. Tonight, though, no one would brave the creaks and cracks that were coming towards us from the darkness. The spinney could wait until daylight before showing us inquisitive youngsters its secrets.

That first day of non-stop activity was brought to a heart-pumping conclusion. We returned to the huts, where the fire's glow was pushing back the blackness, and I met Christine. We were both nine; she was slightly shorter, with fair hair, a permanent smile and, as I soon found out, a knack of making a heart beat so loudly it could be heard over a tractor engine. I'd seen her a few times during the day, but we hadn't spoken. Now, as I wandered towards my hut, she appeared out of the shadows and whispered, 'Dare, truth, kiss or promise?'

I had heard my brothers talk about this game and, although I had never played it, sensing that I was about to made my saliva dry up and disappear. With tongue glued to the roof of my mouth and in the sure knowledge that Christine could see the turmoil I was in, I stammered, 'Kiss.'

She touched her cheek with her right index finger. 'Kiss me here,' she ordered quietly, like Estella from *Great Expectations*.

As I bent towards her she quickly turned her head so that my lips were planted firmly onto hers. The soft, warm sensation caused an incredibly pleasant tingle to spread from my feet to my hair and back. I was rooted to the ground, unable even to blink as she just squeezed my hand and skipped off into the night.

Some minutes later, while I remained in this static state, my eldest brother Pat, a fifteen-year-old whose abilities at winding people up should be recognised by the England selectors, called me. I don't think I responded, because in an unusual show of compassion he strolled over and said, 'What's up, bruv? Homesick?' I was too embarrassed to tell him that completely the opposite was true. I'd just had my first real kiss and far from feeling homesick, I was mind-numbingly lovesick.

In that first year of visiting Petham, an event occurred that was as unusual to the locals as it was to us London workers. For several days varying numbers of army vehicles were driven across the hilltop to the right of the camp. It was quite disconcerting for the youngsters who, with no television and little access to newspapers, had no idea what was going on, while the adults were talking about another war. This gave them an excellent opportunity to replay their part in the last 'little fracas'. It may very well be that some of the valiant deeds spoken of were true, and I have no intention of denigrating our servicemen, but it is usually fair to say that the noisiest faction saw least action.

Anyway, clearly they were not too well informed, because I later discovered that it was almost certainly the ending of the war in Korea.

An armistice was signed in 1953 and the trucks passing across the horizon were apparently all returning from Korea. Anyone watching the American television series *M.A.S.H.* could be forgiven for thinking that the USA stood alone against the North Koreans and the Chinese. There were in fact sixteen UN countries involved in a war which claimed the lives of five million civilians and armed forces personnel.

3

Discovering Petham

Waking up to this new land of Kent on the Sunday morning caused my feelings to veer from excited anticipation to nervous tension. Memories of the previous day filled my mind and that kiss was the most persistent of them. Although I tingled whenever I thought of it, I was apprehensive at the idea of seeing Christine again. Perhaps that would be best left to fate.

The allocation of the huts was reasonable. The two that we were given, and continued to use every year, had the advantage of being opposite the fire and right next to the cookhouse. The males had one, the females the other. We lived at the top end of the camp, feeling very superior because most of the family groups had huts there. We were younger, noisier, went to bed later and probably picked more hops than those ancient bodies in the lower section.

There were nine Lawsons in total, despite which we still managed to get around. We had an old combination motor-bike that I think was a Panther. The sidecar held the youngest five of us, then if we were only going a short distance, Pat, and Tom, who was thirteen, would go on their bikes. Dad came to Petham on that first weekend, the same as most of the men. He made the journey as often as possible, which happily meant he was down to see us most Friday evenings and stayed until late Sunday.

The first night, it was difficult to sleep, not only because of the kiss, but the mattresses were most uncomfortable. They consisted of just two sheets sewn together and filled with straw;

the correct name for them is a paillasse, but it was 2003 before I found that out. We were all awake by 6 a.m., and getting out of bed was like stepping into winter. Ted, who had been here the previous year, knocked softly on the door, whispering that he was off mushrooming. I dressed and joined him. We walked up a steep hill to the right of the huts, across a wide cornfield and onto a common, where a herd of dairy cows moved aimlessly around, a cloak of steam marking their journey as they developed their own methane mountain. The morning dew was heavy and the plimsolls I wore were soon soaked through. Strangely though, the shivering stopped once we began collecting the giant field mushrooms. We gathered two dozen from that one field, then headed back.

It was a little warmer as we made the return journey through the cornfield and we stopped to take in the painting before us. The huts formed a distant backdrop in which a few tiny people could be made out, stacking the logs that had been delivered to us by tractor. Wisps of smoke curled up from the damp wood on the newly lit fire and a slight mist was brushing the fields. The feeling of serenity was dispelled by a baby screaming, which seemed to bring the camp to life.

We completed the distance back in a jog and by the time we reached the huts, we were more than ready for a fry-up. The mushrooms were warmly welcomed and tasted almost as good as they smelt. I probably went mushrooming three or four times each season, but the feeling that was so strong when looking down from the cornfield was never repeated; the magic was all in the first viewing.

That first full day on the farm was a Sunday and we spent the time exploring the countryside. Ted, who was intent upon showing us everything by lunchtime, started with the chalk pits, of which Kent seems to have an abundance. The main problem with them is that although they may be interesting to walk around, and exciting to climb, it took only one fleeting look from the top to confirm that my acrophobia was alive and well. It seemed that the others also suffered a fear of heights, because not only would none of them go near the edge at the top, but no one ever suggested going to a chalk pit again.

Ted moved us on to the orchards, then proudly showed us the village cricket pitch, which had a tree inside the boundary, just as the Kent county pitch has. We saw the manor house, and the old barn, which was to be the source of many pleasant evenings, and ended up in the village. The shop was closed and we were sweetless, so on the way back we scrumped a few apples, which, because we rarely ate fruit in London, were as tasty as the sweets would have been.

Even though he'd only been here once before, Ted knew all that was needed about where the best fruit was grown, the quickest way to get places, where to hang a rope for the best and highest swings, how to make a bow and arrows and which two huts had the little holes in the back in case you wanted to spy on anyone. That first day all of the kids seemed to separate naturally into groups. Ted and I, Arthur, aged nine, and his brother Ray, who was seven, spent most of that season together – and that often spelt trouble. From that first Sunday, when we discovered a wasps' nest by the standpipe, we seemed to attract problems.

Although it was already afternoon, none of us had yet washed. The four of us took our towels down to the bottom end of the huts, where the tap was. We had started to wash, and begun splashing each other, when we noticed a few wasps circling almost at ground level. We tried to shoo them away, but they appeared to double in number. Like a magician's trick, we shooed them again, with the same results, until we were facing more than we could count. Arthur, calling on his renowned ability to take a calamity and turn it into a catastrophe, picked up a stick and began to thrash it around until the wasps were in a frenzy. Too late now for us to walk away whistling, they were suddenly in a stinging mood.

With the exception of Arthur, we all ran in different directions. The shorts and the vests we wore gave little protection and much later, when we compared our sting tally, we found that Ray had been hit about a dozen times, probably because he wasn't a very fast runner, and also he had just headed for the field. It wasn't until his dad caught up and began swinging his coat around like a helicopter blade that the wasps dispersed. Ted had headed straight along the front of the huts, so the wasps that were initially after him now began stinging the people who had been relaxing by the fire or were in the cookhouse. Unfortunately, I missed seeing the mass exodus from the camp, but apparently a lot were stung, and I imagine a few of the bigger ladies, especially Doreen Kerslake, who weighed over twenty stone, required oxygen by the time they completed a lap of the field. Ted had stopped as near to the fire as he could and presumably the wasps turned away from the heat, or forgot him, as he only suffered two stings.

Arthur didn't go anywhere. He just kept on thrashing with his piece of wood, then tried to crawl under the tap to soak himself. Maybe the water made him so cold that the wasps, after one attack, left him alone. He agreed that the seven or eight stings he did get didn't hurt anyway because he was just frozen. His teeth were still rattling an hour later.

While this was happening I just sprinted for the woods. That's lucky, I thought as I passed the rope swing that Ted had put up that morning, I jumped up, grabbed the rope and ran up the hill with it. Then, with a breakaway group, probably led by Biggles wasp, aimed at my head, I launched myself forward. As I reached the furthest point out, about twenty feet from the ground, I was still aware of the wasps around me. Then the rope, with my weight on it, came undone, I flew forward another ten feet and, still holding the rope, fell some distance down into a huge bed of nettles, I lay there for a while, realising that I had lost the wasps but gained an all-over nettle rash and an assortment of bruises, I waited a few minutes to see if they would return, then I hobbled back to the huts, picking dock leaves as I went.

That was not the only swing in the area. Ted had shown us a couple of others that he had put up, and his tree-climbing and rope-knotting was better than most. He hadn't made such a good job of the one I had fallen from, though, as my battered body would give evidence to. Actually, that became part of the fun. The rope was not always attached tightly to the branch, and the people using it could not be sure it would hold their weight. If it didn't, there was the possibility of a world record long jump, often into a carpet of nettles.

As the season went on, the number of minor accidents increased. By October of year one, Arthur had the longest swing-assisted jump and was totally unhurt. Eric Pearce had the longest, and loudest, scream on landing, running what must have been a record time back to the huts and the warmth of his mummy's cuddle. You could hear his cries bouncing up and down as he ran, and if his dad had been there he would have received a whack for being a cry-baby.

Towards the end of that year and those following, either Tom or Pat put up the ropes. Although there was still the occasional accident, that was only when wear and tear caused the rope to snap, which only happened once or twice a season. Incidentally, I don't know if Kent was then the champion nettle-growing county in the UK, but if it wasn't it would certainly have been very close.

Later on in the day of the wasp attack, I was wandering around the camp, not willing to accept that I was looking for Christine. I hadn't seen her for some hours and as the fires were burning low and people were drifting off to bed, I assumed that she had turned in for the night as well. I glanced into the cookhouse, where the embers of a fire sparked and died. A chair remained by the hearth and as I walked over to

14

sit down, there was a brief and final spurt from the fire, which lit up the young girl who stood behind the door, smiling. After the shock left me, I asked her what she had been doing that day. She walked slowly over and sat on my lap, luckily, it was too dark for her to see how much my cheeks burned, but she probably felt the heat from them. We spoke for about half an hour. I told her about my battle with the wasps, she said she had heard what had happened and I must have been very brave, though she smiled as she said it. My mum's voice then cut into the warm sleepiness: 'PADDYTOMMYNITABOB-BYEILEENRICHIEBARBY.' She was directly outside the cookhouse, calling her flock, and I squirmed. Christine slid from my lap, gave me one kiss that kept me seated, and left. I heard her say goodnight to my mum, then I took a few deep breaths and strolled out as if everything was perfectly normal instead of it just being perfect.

It was difficult that year to spend much time with Christine as I enjoyed getting into unintended scrapes with my mates. We did get together sometimes in the evening, but best of all was when we were all there, her friends and mine, and we went over to the old barn, which was filled with bales of straw, and played what was now my favourite game, dare, truth, kiss, or promise.

4

Why We Are All Here

The actual work of picking hops began the day after the trouble with the wasps. A tractor, with a trailer on tow, came to the huts about 7 a.m. Everyone clambered on board with their chairs, baskets and food and drink for the day; everyone, that is, except Mrs Kerslake. The tractor had a large shovel attached to the front, which had a chair on it. The shovel was lowered to the ground, Mrs Kerslake sat on the chair, and it was raised about four feet into the air. I've no idea how she got to the field unscathed. The chair bounced about as if it were on a trampoline; it was amazing that Mrs K wasn't seasick at the very least.

After taking us to the first field to be worked, the driver arranged to return at 5.30 p.m. We all got down from the trailer and a scene resembling the Klondyke gold rush began. Each family or group had been allocated a lane, or lanes, to work down, but because some of the more experienced workers thought that some rows were more productive than others, they had already walked the field the previous night and mentally selected the rows they wanted. They now ran to them like the proverbial claim-staking miners. It almost seemed to be a tradition, but as we knew nothing about it we actually stopped and looked on in astonishment as arguments began.

By 7.30 a.m. everyone was sorted out and had started working. The way to pick hops, as explained to us by Aunt Marge, went thus:

16

First, pull on the bine that, like any climbing plant, had grown up a length of string that was attached to the wires above. This should bring down the complete bine. If any remained on the wire, a guy called a polepuller was called for, to hook or cut it off. The bine was laid across the lap, bushel basket at the side of the chair, and the hops were picked into the basket, either individually, which was very slow, or by first removing the leaves, then stripping the hops from the branches with one sweep of the hand.

The owners were the three Collard brothers. One of them would come round three times a day with the farm manager and a worker to collect the hops that had been picked. The full tally baskets were tipped up into a sack and checked that they were 'clean'. Apparently it had been known for pickers to try to put a whole bine in the basket and cover it with a few hops. The three wise men heralded their arrival by shouting, 'Tally up, all full baskets.' There were five bushel to a tally and the main person in each line was given a notebook in which a record of tallies collected was kept. The bulk of the money earned was left with the farm owner and paid out in a lump sum at the end of the season. Most workers had a sub each week that just covered their living expenses, with perhaps one trip to the Petham Arms.

I don't know what the children were paid, I think most of us got just a shilling or two each week, then two or three pounds at the end of the season. No one minded; whatever we received was more than if we hadn't come, plus we finished early most afternoons, providing we picked one tally each, so that we could go off having fun, and double plus, we weren't going to school.

One end of season, mum must have had a good year. She gave me six one pound notes when we left. I cut about fifty pieces of newspaper to the same size as the notes, then put three pounds each end of the bundle and an elastic band around it. I then went with Ted to the shop off Edith Road and flashed the wad in front of the lady who worked there and a person she was serving. The customer's eyebrows were lost in her fringe as she saw what she assumed was a pile of money. The shopkeeper asked where I had got it, and said she would call the police if I didn't tell her, so I showed both her and the customer my *Daily Sketch* cut-outs. They both had a go at me for being stupid and asked if anyone else had seen the bundle. They hadn't, luckily, and I realised that the women had been right to call me stupid. After all, what sort of a bodyguard would Ted have made if someone had decided to rob me. Ever since that day I have always been aware of the chance of being mugged and I rarely even take my wallet out where people can see in which pocket it is kept.

Anyway, I thanked the ladies and offered them a toffee each, and giving away sweets was not something that occurred very often in my childhood, I had a theory that it was sweets that kept people healthy. By way of proof, Pat didn't eat many sweet things, he liked his teeth too much. Is it just coincidence that the only one among us who seemed to suffer any side effects from the hops was Pat? They sent him to sleep, as if he was suffering a bout of narcolepsy. Three or four times every day his head would slowly sag forward until it was hanging just above his knees. We sometimes left him like it so we didn't have to listen to him griping about the smell, the taste of it on his food and the fact that the hops stained his fingers green and it could only be removed with elbow grease and a pumice stone. The rest of us not only stayed awake, we ate Pat's sandwiches, hop stains and all, while he slept.

Of course it was a fact that Pat would sometimes awake, after a Sleeping Beauty turn, in an awkward position. The feeling would disappear from large parts of his body, rather like when a dentist numbs his patient's gums. He would then make peculiar noises until he was noticed, and get someone to knead his muscles to start the blood moving again. Once, he fell into one of these coma-like snoozes and somehow crossed

18

his legs in such a way that both of them went completely numb. As he went to stand up, he fell in a heap on the floor. He continued to do this until he regained some mobility in his limbs, then of course entertained us further by suffering a double leg full of pins and needles, which is akin to being bitten repeatedly by a horse fly. He stamped his feet on the floor, alternating between one foot and two, like a Red Indian war dance, then sat down rubbing his knees, asking for lunch.

Pat was only there the first two years. He rarely came down at weekends, such was his dislike for Petham, but in the second year he was given the chance to outrun a young lady for an evening in the barn. He had met Sally a week or so into the season – she was in the red huts for that one year – and had been going out with her for a few weeks. Things were going well except Pat, being a normal healthy teenager, wanted more, like an evening in the hayloft with dare, truth, kiss and anything goes. She led him along a little then said they would have to split up, but she first made him a very unusual offer. They would have a race from the black huts to the village; she would have just a ten-yard start, and if he caught her he could take her to the barn that evening and she would be putty in his hands, so to speak.

My brother developed tics in several places, asking what time the race was. Sally decided that now was as good a time as any and they made their way to the ditch at the bottom of the huts. They had just reached it, when Sally shouted 'GO'. She ran loose and even, with a stride that took her thirty yards clear of Pat before they had reached the manor house. Seeing the shapely behind drawing away from him, and imagining an evening in the barn with her, made him increase his tempo. His legs were now going like pistons in an effort to overtake her and for a while he was actually catching up with her. As they passed the bull field, however, which was about halfway, she glanced back, waved at Pat, and increased her speed so that she just cruised away from him. Needless to say, she was sitting outside the pub when Pat arrived a few minutes after her. She asked him to walk her back to the red huts and on the way told him she belonged to an athletics club in London, her distance was a mile and upwards, so the race they had just had was like a short daily training run for her. When they reached the huts Pat gave her a goodbye kiss, then asked her, with a plea in his voice, whether she had ever lost one of these 'special' races. Sally gave him an enigmatic smile and walked away.

One of the first jobs in the morning at the camp was to gather kindling wood for the fire and it seemed to fall upon whoever was up earliest. The fire was usually lit by 6.30 a.m., earlier if someone had been mushrooming. Tom was often out 'finding the fungus' and one Saturday as he returned, he saw that the gypsy family, who for this season had been living in the top hut, had a good fire going, whereas our usual firelighters were either still in bed or creeping around, severely and deeply hungover. So Tom collected the kettle, filled it with water and took it up to put on the gypsies' fire. No one spoke to him, they just watched him whisk it off as it boiled, take it back to his hut and make a pot of tea for the family. The next morning he saw the gypsy fire was again roaring and whoever was trying to light our one clearly was unaware that matches had been invented. Tom went up to the gypsies' fire with a kettle of water again, but this time a couple of gypsies confronted him. They made sure he was aware of the difference between ours and theirs and pointed him in the right direction for his huts, only this time he returned a little more rapidly than he went. Tom was not overkeen on the gypsies after that and I think he would have liked to have got some revenge for his damaged pride.

One of the families near us had a ritual with their food, but only one particular thing, eggs. There were ten of them, almost all were girls. Once a week they brought in hard-boiled eggs, plus one that was raw. At lunchtime they each cracked an egg on another person's head, leaving one of them with their hair in a sticky mess. It seemed to me that it would be easy enough to tell the difference between a hard-boiled egg and a raw one, and anyway, wouldn't the ones that were boiled smell a bit 'eggy'? But perhaps it was a new way of conditioning.

As the days passed and we got to know the area, we realised from all of the steep hills on the local roads that it would be handy to have some form of kart to pull or push around, both for fun and for carrying shopping, or scrumped fruit. This thought happily coincided with finding a set of pram wheels near the gypsy caravans. When we asked about them nobody could tell us who the owner was, so we took them and, using some old planks we begged from the tractor driver, we assembled a rickety kart, or barrow as it was called then. The first day we used it, Arthur was holding on to the string that should have steered it. He came round the bend with a look of sheer terror on his face and the nut that held the front axle in place fell off. It had only been screwed up finger-tight because we had no spanners. The front wheels continued down the hill on their own and Arthur was thrown onto the grass verge. After running repairs and a refusal from Arthur to ever get on it again, the answer seemed to be stone, paper, scissors to decide the next driver. Unfortunately, I won. We pulled the kart back up the hill and, not wishing to appear chicken, I got on. Ted and Arthur were on each side, with Ray at the back, and the three of them clearly looked forward to my terror-stricken pleadings for them to stop. My stubbornness, however, was stronger than my fear, at least for the first ten yards. The three pushing gave it all they had and by the time we had travelled fifty yards, they couldn't keep up with the kart. I managed to lie down close enough to the ground to clear the bend, but the kart was rattling so much, the bolt dropped out again and the whole thing collapsed.

I was thrown forward as most of the kart stopped dead. My elbows acted as brakes to stop me on the gravel-covered road. Both of my arms were indented with little pieces of stone, probably left over from the winter gritting. The burning pain had me on the verge of tears, which I overcame by yelling at the other three about what I was going to do to them once I had received treatment to my arms.

Back at the huts, my mum picked the stones out with a pair of tweezers, like removing buckshot, then filled the holes up with iodine. Probably no research has been carried out on how much more pain a man can suffer without blubbing if his mates and girlfriend are watching him being repaired. I told them it was the iodine making my eyes water, but a mixture of laughter and coughing followed me as I sadly limped from the hut and went off to be alone. It was not until next morning that I realised there was an upside to the crash. As I was unable to bend my arms without pain, I was not able to pick hops. I spent most of the week strolling about wearing that air of resigned suffering, and accepting the odd sweet and words of sympathy, which all helped to heal the wounds no end. The scars on my right elbow still remain half a century later, but I never discovered what happened to the remnants of the kart; they disappeared without trace.

After living like one huge family for so long, we were now within a few hours of separating and travelling to different parts of the country. The hop fields had finally given up the last of their crop and now resembled a wasteland. The camp appeared almost deserted as everyone packed away their belongings to await the removal men.

On the previous day we had all given a cheer as the last of the bines had been pulled and stripped. After lunch most of the workers formed a queue around the manor house and were paid the money that they had amassed over the last six weeks. The Lawson pay-out was around forty pounds, a very useful sum of money in 1953. We might have increased this by fifteen per cent had all of the family managed to stay awake at their baskets.

This pay-out routine made for an enjoyable few hours, with people dancing and singing and, for a short while, feeling rich. That evening the camp almost emptied as most of the adults had a few drinks at the Petham Arms. Many of the children, even the young ones, waited outside the pub, and the parents sent out crisps and lemonade for them. That night before leaving must have been the most lucrative of the year for the publican. Not only were there the locals in, but also the black huts, the red huts, the gypsies and most of the staff from the estate.

The joy that had overtaken the workers on Friday was tinged with sadness on Saturday. The hangovers melded with the unhappiness of packing, and the knowledge that we would be back in London at our schools and jobs in the next two days. Just before the lorry arrived to pick us up, I had a last walk around the huts. Most of them were empty and had a look of loneliness about them, after the continual noise and movement they had endured over the past few weeks. As I made my way to the back of the camp and looked across the desolate fields and threatening spinney beyond, a small hand gripped my fingers. I looked around at Christine, gave a watery smile and we walked back without talking, to stand by the remains of the fire.

It was a downcast group that the removal men met that October, completely different from those happy souls they'd dropped off in August. Even the youngsters who'd met the lorry at the gate and raced it back lacked enthusiasm.

The loading cheered everyone a little and within an hour we were almost happy to be going. As we drove across the field, we waved to the gypsies, who were also packing to leave, had a last view of the hill where the swing still trembled a little in the breeze, turned uphill towards Chartham Hatch and were homeward-winging – singing.

5

Pages of Pain

The second year that we worked in Petham had been anticipated for weeks beforehand. There was a build-up of excitement amongst us all, especially since we had already been off school for four weeks of the summer holidays. This enabled us, Ted, Arthur, Ray and me, to get together a couple of times before the season began. We sometimes cycled from Edith Road, Leyton, to my house at 90 St Ann's Road, Tottenham.

We had moved into number 90 in 1946, but I am unable to remember anything clearly until my first day at Stamford Hill Junior Mixed. I was a week over five years old and I had been looking forward to going, as Pat, Anita and Tom attended there and I knew they all enjoyed themselves.

The teacher hustled us into line and I was obviously in the infant section. We all took our coats off and shuffled into the hall, where we were assembled in lines according to class groups. We had just stopped moving when the girl behind poked me and just said something like, 'Hello, I'm Carol.'

Immediately one of the teachers strode over to me, lifted the leg of my shorts and gave me a ringing smack that echoed round the hall. Everybody seemed to look over, but although I felt like crying at the unfairness of it, I wouldn't let that old cow see she had hurt me. I suppose like other children, when on the verge of tears, my initial reaction was to poke my tongue out and say, 'Didn't hurt.' However, I was wise enough not to stick my tongue out until the teacher had turned away. I noticed as I did it that one of the other teachers was looking at me. I thought I would be in deep trouble, but she just shook her head slowly. Perhaps she didn't like the old cow teacher either. When we filed into our classrooms, one of the girls in my set was on the table, dancing around holding her frock up, then she pulled her knickers down just as our form mistress came in. It was the lady who shook her head. I think I was already in love. She quickly got the girl off of the table and, acting as if nothing had happened, began calling our names out. When she got to mine she had to repeat it as I was just sitting looking at the red hand print on my leg and thinking that if I came back tomorrow it would only be because of our form teacher. I found out later that she was twenty-one, had just begun teaching and within the first term every boy in the school would be besotted by her. She would also be married the following year.

Before then, however, just a few months after I had started at Stamford Hill school, I suffered a most embarrasing hour or so with her. The class were all in the hall for one of my favourite lessons, country dancing. I always joined in with an excess of enthusiasm and was, with my partner, ducking under the bridge the couple in front had formed, when my trousers shredded. I had inherited them from Tommy to wear at my first school, but they appeared to be shrinking each week. The legs were now tight enough to restrict the circulation, so although there was no chance of them falling down, a split from belt to trouser bottoms still exposed enough of me to have the class in hysterics.

The teacher took my hand and quickly ushered me to the classroom, stopping on the way to arrange for someone to take over the dancers in the hall. When she was seated she opened a drawer and held her hand out. I looked blank. 'Trousers.' There was a definite edge to her voice. I walked slowly to the opposite side of the desk, took off my trousers, and handed them to her. She looked at them in amazement, then after a few minutes trying to sew up the split, she dived back into the desk and emerged with a large red patch. Explaining that they were too worn out to be repaired any other way, she began sewing with gusto. Sadly the class arrived before she had finished. They filed quietly to their seats, then the ones just behind me began to laugh. The teacher walked around to see what was amusing them, and my face was burning. The tail of my shirt had been cut off to use as a handkerchief, which was not totally unheard of, but the fact that I never wore underpants was the cause of all the merriment. The teacher went to a small side room and came back with a coat, which she gave me to wear while she completed my trouser repairs. As soon as she had, she gave them to me, told me to put them on in the small room, to leave the coat and to go home. The other children created a hubbub, saying it wasn't fair, why should he be allowed to go early, and so on. I went over and thanked her for what she had done and the smile she gave me made my chest ache. I turned to leave and a great cheer went up from the class as, for the first time, they saw the red patch *in situ*. With a regal wave I exited stage left and silently vowed never to wear those trousers again: they could be handed down to Richard.

Returning to 90 St Ann's Road for a moment, it used to be a terraced house on three floors. The German air force had restructured it into a semi-detached on sloping floors. We had the upper two floors, which comprised a dining room, lounge, kitchen, toilet and one bedroom on the first floor and two attic rooms on the second. The house to the right as you face them had been hit by a bomb during the war, and seemed to be in a fairly dangerous condition. That did not stop Tom from climbing out of the second-floor landing window, crossing the roof and shinning down a drainpipe so that he could ride around on Jim the window cleaner's bike. He clearly didn't see any danger. Jim had the key to the big double doors that led through to a garage, where the bike, the little box sidecar and the ladder were all kept. Tom said he would pretend to be a motorcycle rider when he was on the bike; in fact, as soon as he was old enough he immediately bought his own motorbike, a 350 BSA which cost him thirteen pounds. He still recalls the registration number, ENO 999, which sounds as if it's the emergency services for chronic indigestion. That was the bike he helped move all our possessions to Harlow on, and with dad's combination, plus Pat's recently acquired Triumph 600 sidevalve, they made a formidable trio. The most impressive was clearly the combination that dad had actually strapped a double wardrobe to. With all of the other pieces of furniture and household items that were tied to him and the motorbike, he only needed a lance and he would have made a superb Don Quixote attacking the windmill. Paddy's Triumph was constantly blowing the head gasket, due to a cracked cylinder head, so he always carried his own spare piece of gasket, which enabled him to carry out on-the-spot repairs if needed. He apparently became very efficient at whipping the head off and replacing the parts needed.

To return again to St Ann's Road. After Tom replaced Jim's bike, he would climb back through the window into No. 90. The rooms there were superb for playing in. The floors were at such a tilt that putting down anything with wheels on meant it would just roll across the room. The stairs were at a twenty-degree angle and everything creaked. We lived above Mr and Mrs Gumbritch, their son Dennis, and for some of the time their daughter and grandson, whose names now elude us all. According to one well-paid grass, whenever the child was staying there, Tom used to slip into the kitchen downstairs and pinch a few spoonfuls of the National dried milk powder that was kept there for the baby. No wonder he outgrew his shoes so quickly.

With nine of us constantly making a noise, it must have been a non-stop bundle of fun for them. They did seem to spend an awful lot of time banging on their ceiling with broomsticks, and sometimes that would be answered by nine pairs of feet leaping up and down in unison until it seemed that the floor must collapse.

At that time dad was driving for a company called Lebus, and Tommy used to miss school sometimes to go with him. Articulated lorries had a top speed limit of only 20 m.p.h, and his longer trips, to Wales or the north of England, could mean him being away for days at a time. Initially the idea of staying out overnight was appealing and Tom went whenever he got the chance, but boredom inevitably set in and he would end up finding out the distance between telegraph poles, then counting the poles to work out how many miles they had covered. Occasionally dad's return would be in the early hours of the morning. So those Gumbritches not only had to put up with the possibility of a ceiling collapsing, the noise of seven children constantly running around and four bikes being left in the passage, they also had a disturbed night once every couple of weeks. But we didn't feel any remorse; they would often lock the back door when they went out so we couldn't get into the garden and they were forever moaning about the number of times we did go out there.

Unfortunately, to get to the garden, where the washing line and coal bunker were situated, we had to use the Gumbritches' back door. One particular time, dad had argued with Mr G and although we needed some coal, he was too proud to ask if he could get out into the garden. He searched for some string and, unable to find any, he collected an assortment of shoe and boot laces from us, tied them together, then attached them to a bucket. Pat jumped into it and was lowered to the bunker. He half filled the bucket with coal, but unfortunately, as dad began hauling it up, one of the knots must have come undone and the bucket fell back, luckily only a couple of feet, on to Pat's head. I don't know if he required any treatment, or has any lasting scars, but dad was quick to run down through the now unlocked back door and tend the egg that had so recently been laid. As a result of this incident the washing line was now amended. The ends were joined together, to form one long loop, so that mum could now hang out the clothes from the window, just by pulling the line along until the next space in the washing appeared.

Almost opposite 90 St Ann's Road was a grocer's shop and its milk was delivered early to a side alley in Paignton Road. If we were short any day, Tom would wait for the delivery lorry to go, then jump over the gate and liberate a couple of pints. He used to call on the greengrocer two or three times a week to buy three pennyworth of 'specs'. The shopkeeper would fill a paper bag with fruit that was damaged, and Tom would just cut the bad parts out. The baker was also approached for a pennyworth of stales; he would generally top a bag up with stale buns or cakes. He knew most of the food outlets in Tottenham, did Tom. I reckon that the distance he walked following this quest for provisions was the cause of his non-stop need for cardboard to put in his shoes and save his feet from wearing out.

Talking about walking, mum had a friend called Stella, who lived in Edmonton. She used to visit her at least once a week, sometimes with two children in tow, but usually three or four. The distance amounted to a round trip of about eight miles. Mum would borrow ten shillings from Stella on a Tuesday and return it on Thursday when dad got paid. The journey was made all along Tottenham High Road, which was usually packed with shoppers, so the going was slow and the big old pram with a child either side took up most of the pavement. I enjoyed those trips to Stella's, mostly because at her house there was a garden we could play in. She also had a daughter, called Chrissie, who was the same age as Tommy and I think they had a little thing for each other. Anyway, I do remember that whenever I met Chrissie I always wished I was a few years older.

* * *

Dad bought a few different vehicles in his time, but I think the one remembered most from all the motorbikes, combinations, cars and vans, was the old London taxi. It had a pull-back roof and looked as if it was being used in a *Carry On* film. He bought it just before Christmas and to try it out we had all gone over to Uncle Jim and Aunt Marge's in Leyton. On the way we had three people flag us down, thinking we were working. When we got there the adults decided to go for a drink. Marge had made us a lovely-looking Christmas cake and mum suggested we put it in the taxi to save us having to take it at a later date. So the cake was put on the back seat and they went off to the pub. On the way, apparently, they had pulled up at some traffic lights. A rather inebriated man, thinking the taxi was for hire, shouted out some drunken sentence, opened the back door and jumped in – straight onto the cake, which crumbled into many pieces. The drunk, realising that something wasn't quite right, jumped out and staggered off, while the rest of us, once they returned from the pub, got on with what was now the most important job, to clear up all the mess, I suspected that wouldn't take too long as we all began eating, our chewing laced with sounds of appreciation.

Food was one thing I was unable to concentrate on for a while earlier that year. Tom and I had spent the day at Clissold Park, one of the many local areas where we could go with a football and a packed lunch and just stay there for hours. We would usually meet up with other youngsters and get a game going. This particular day as we made our way back from the park, I stupidly climbed up onto a garden wall and began to walk along it. Unfortunately, I slipped and fell, biting my tongue quite severely, and as Tom said, 'Boy, did that bleed.' He got me home as quickly as possible, but I was covered in blood, which must have been a shock for mum. Still, she took over, mopped me up and got me to the doctor,

31

who somehow stemmed the flow. I'm not able to recall the name of our doctor, but I know his surgery was a house in the top end of St Ann's Road and anyone attending for the first time was usually quite interested in the large metal table that took up a substantial part of the waiting room, Apparently it was designed for the patients to squeeze under if there was an air raid warning during the war. I think that, had they been able to carry it, an awful lot of those patients would have had that table down to the scrap yard before the doctor could call, 'Next please.'

The nurse who was in the surgery appeared very efficient and I suspect that because she couldn't put a dressing on my tongue, she felt obliged to do something else, so she drew off some O negative for a blood test. If I'd been donating blood that week, the needle would have been frantically searching for a vein that hadn't already been emptied.

On the day that we travelled to Kent for our second year of hop-picking, it seemed that we were in Dr Who's Tardis; it was difficult to believe so many people and so much furniture and bric-à-brac could be fitted into such a confined space. Meeting everyone again at Edith Road, it seemed we had never said goodbye. This applied especially to Christine. Apart from an initial shy smile and self-conscious wave, which was not her reaction last year, we continued together without any awkwardness at all.

The trip to Kent went without a hitch and we were once more grubbing up nettles, unloading furniture, lighting fires and eagerly looking forward to exploring the area again.

At the back of the huts was one of the biggest hop fields, The Banjo, which led to the spinney. We had been through it a couple of times, but now we were all a year older, we tried to pretend we were also a year braver and thought we would kick off the new season with a good practical joke. Most of the adults, rather than walking to the pub along the road, took a short cut through the woods. They always went on a Saturday evening, especially on the day of arrival. Our idea was to get to the spinney before them, make sure we were well hidden, then indulge in some scuffling, scratching and growling, hopefully making them so nervous they would run from the woods. That would be fun, we thought.

We reached the outskirts of the woods at the same time as dusk did. Dusk, however, was more frightening than us, so we all stopped for a few moments to recover our courage. Once we were in the spinney, though not too far in, we decided not to split up, but find a place we could all get some cover together. We had found a suitable area surrounded by bushes, just as the rain began. We pulled branches over ourselves, but the rain increased in its intensity and the shelter that we had so hurriedly assembled was useless within minutes. To add to our worries, a bolt of lightning seared through the darkness and for a second lit the woodlands like Oxford Street at Christmas. Without a word we all ran at once for the path that led out of the woods. The thunder that crashed through the trees seemed to be attacking us, and as the rain hammered against us, we pumped our legs as fast as they had ever travelled and within five minutes we had reached the huts. We fell into the cookhouse, thinking that if it was struck by lightning, it wouldn't go up in flames as the huts might.

When our bodies had defrosted and the rain had almost stopped, we went out to relight the fire and find out where all the adults were. There was a taxi in the village, an old black Austin 16, but no way of contacting the driver, unless someone was prepared to go into Petham to book it. That had not been necessary today, though. We discovered that one of the gypsies had given them a lift in his dormobile, so we thought that they would have a long walk back from the village in the cold damp air. But then the twelve-seater came round the end of the huts with at least twenty people squeezed in it. They were hanging out of the sliding doors and windows, singing about friends and neighbours and scaring the wildlife for miles around. The driver, who none of us knew, seemed to be plotting his route by the stars, swerving, bumping and generally guessing when to brake. His passengers were all so inebriated they didn't know whether he was driving or they were. Still, he parked near enough to the fire that the people getting out through the side door soon sobered up.

The driver went with the others to one of the huts for a nightcap. Apparently he then wanted to move an oil lamp that had been alight for some hours to another part of the room. He picked up the lamp by the glass top and never noticed it burning his skin, but fortunately someone did and they took the lamp off him, saving him a certain trip to the hospital.

Next morning he came over to say thanks for the drinks, then casually asked if anyone knew how he had blistered his hand so badly. It didn't seem to be hurting him, which indicated that sobriety was still a few hours away. This thought was reinforced when he took a hip flask from his pocket and drained it.

I never saw any police on traffic duty in the Petham area, but if there had been one that night, and breathalysers had been common place, that man would have been both in the record books and in prison.

Alcohol was one of the few items that had to be paid for. We could scrump as much fruit and veg as required, and had access to chickens' eggs, wild nuts, two types of mushroom and an abundance of blackberries. We also, for a short time, had some very tasty rabbit stews, casseroles and pies. Unfortunately, in the early fifties, myxomatosis, a disease introduced to the rabbit population in the UK and Australia as a pest control measure, almost wiped them out. It caused the eyes to swell and the eyelids to close up so that the rabbit or hare was blind and feverish. Eventually, unable to move, it would just die. There were many days that thirty or forty rabbits could be seen, just sitting and suffering. Some of the men, and older boys, would club them across their neck or head to end their misery. That was one of the few sad things to happen in the first few years at the farm.

This second year of hop-picking made me feel a little more adult, because I had this knowledge about what we did and places we went that I knew no one back in my school had. I'd even given a talk on hop-picking to my class. Sometimes I felt I could even have tackled Doody, the teenage gypsy girl. She had all the young boys dreaming of her. She was there again this year, and had never spoken to anyone who wasn't in her group. The gypsies rarely had anything to do with us, despite living less than a hundred yards away and working for the same boss.

Another thing they had nothing to do with was our toilet system. They probably had smart new ones in their caravans. We had two for all the people living in the huts. One of these was a black wooden structure, with a single seat attached to the walls. It was like a small version of the huts we lived in. It was about thirty yards from the camp and near to it was a two-seater de-luxe in red with a corrugated roof. The seats comprised two holes in a plank, stationed over what was known as a 'stench trench' and attached to the main structure. When the hole was approaching half full (or half empty, depending on your outlook on life), two men from the farm would come to change it. They pushed a piece of two-by-two into the slots on either side of the shed, lifted it like a rickshaw, then walked it to a pre-dug hole and manoeuvred it into position. They did the same with the other toilet, using the earth they had dug from the new holes to fill in the old ones. Unfortunately, on one of these change-overs, they did not check whether anyone was using the ladies. Mary, four foot ten in stilettos and about eight stone in weight, was clearly not singing, whistling or drumming on the seat as most of the women did when occupying that particular space.

She must have been shocked speechless as the men, not aware she was inside, lifted her three feet in the air, then with a little moan about the weight of the loo, walked her, still in the sitting position, to her new destination. Sadly, they were not able to get her there quickly enough and a trail of number twos, probably brought on by the stress of moving, led to the resited toilet. The workmen, unable now to miss the sights and sounds of the barely coherent woman, nevertheless acted as gentlemen would. They continued as if nothing had happened, placed the shed, and also therefore the passenger, into exactly the correct position, and went back to fill in the old hole, stopping briefly on the way to shovel a few loose turds into the hole they were to fill.

Mary didn't surface from the loo for at least half an hour. By the time she did, everyone had got to hear of the event and she was applauded all the way back to her hut, her face glowing so brightly she would not have needed to light her lamp that evening. It was unfortunate that Mary happened to be the one in the lavatory when it was moved, because there were not many women in the camp that the men could have lifted quite so easily. There were also few that would have been so embarrassed, which was probably the reason she never returned after that season.

There cannot have been many days in a year when I would have preferred to be in Tottenham than in Petham, but the day the fair arrived each year was one of them. It took up a huge field by the side of the River Lea. All the children went, and usually returned laden with balloons, goldfish, candyfloss and tales of which rides were the best and what was new there since last year. We always had an excellent time, but that year, as usual, most of us got separated. Tom and I decided to head for home.

We came to the last stall, a roll-a-penny which involved pushing a coin down a short wooden slot onto lines of numbers from 0 to 6. If a number was completely covered by the coin, the player won that amount. As we walked past, the owner called me and asked if I would like to look after his stall for fifteen minutes while he went for a coffee. I told Tom I'd catch him up, went into the middle part of the stall and began talking to the passers by. The man gave me some pennies and a small bag and left. I felt a little nervous until a few people had rolled their coins and lost. Fifteen minutes later the bag was nearly full, and I'd only paid out a couple. Then Pat, Anita, Eileen and Richard came by. 'Lucky people wanted, roll a penny and win a tanner.' Eileen looked up dumbfounded. I explained that I was still waiting for the owner. Pat stayed with me, and the rest left for home. When the man finally returned he gave me twelve of the pennies from his bag and I almost turned a cartwheel thinking of Mrs Trett's sweetshop. I now knew what I wanted to do when I left school. I wanted to have my own roll-a-penny stall.

6

Red Huts to the Rescue

We had been in Petham for nearly two seasons before I met any of the workers from the red huts. These homes were slightly bigger than ours, but there were only eight of them. They had a brick-built cookhouse situated at the top end of the huts. The other facilities were the same, with a cold stand-pipe at the bottom end.

Arthur and I had been to the village for some groceries. It was around 4.30 p.m. and as we were approaching the part of road that ran alongside the red huts, a bus pulled in at the stop about thirty yards in front of us and five or six boys and two girls got off. Judging by the way these local kids spread across the road, we were in for some trouble. We both automatically picked up a thick stick and carried on a bit more slowly. The group stopped and began making loud remarks both to us and those around us. Thieving London rubbish, would be the most polite translation of any of them.

The boys were quite different from us in both age and size. The two biggest ones and the girls must have been at least thirteen, the others were perhaps eleven. The older ones were laughing and pushing the younger boys on. We carefully put down that evening's tea and for no reason clashed our sticks together, like sword fencers, then turned to face the locals. As we both discovered several hours later, we had individually considered running off, but admitted we would not have been able to face the others when we got back. Better to collect a beating now than to no longer have any friends.

We had just begun to move towards them, slashing our sticks across the front of us, when, through a hedge behind the school kids, stepped a mean-looking, crop-haired lad of about fourteen. He was followed by a slim, dark-haired boy of the same age, who would have looked academic except that his nose had been broken at least once, which made his spectacles tilt up on one side. They both had clubs in their hands and looked as if roughing up people was something they enjoyed doing. We realised that they had come from the red huts, and with them at the back and Arthur and me at the front, the odds had dramatically changed.

As we turned and walked back, the Petham boys must have thought for a few seconds that we were leaving. A hesitant cheer went up and they made as if to chase us. Then we picked up our tea and turned back to face them. The red hut saviours were slowly catching them up and smacking the base of their clubs into their left hands, giving them the chance to see what was going to happen to them if they didn't run. The locals were baffled that they hadn't been attacked and some fear of the unknown gripped them. They made for the gap between Arthur and me and ran as fast as they could, leaving the girls to follow them in tears.

They had a few catcalls and hisses aimed at their retreating backs, then we turned to profusely thank the red hut lads, who, we discovered, were Mike and Alan from Stratford. We went back with them and chatted by the fire for an hour. There had been a few problems, apparently, with the locals and the red hut crowd. We were too far away for them to bother with, or maybe there were too many of us.

Luckily, we were not involved in any violence that day and perhaps the villagers were embarrassed by what had happened, because there was never any trouble after that between the locals and the London crowd. Strangely, the only person from the black huts who got friendly with the youngsters from the other camp was Arthur. I think that might have been partly to escape from his brother for a while.

The run-in with the locals wasn't the only confrontation that could have led to violence that year. Eric Pearce's father, Barry, was a martyr to indigestion, and on most nights, a short while after chomping his meal down before the rest of his family had finished using the salt and pepper, he would develop a pain in the chest that could only be cleared by belching like an opera singer, loud and long, which left him rolling on the ground, clutching his chest and moaning for a doctor. I thought it was funny, but not everyone agreed with me. Brother Pat, for example, was not one to ignore anything that annoyed him, and Barry Pearce's rain dancing came in to that grouping. One evening, as the squeals of pain began, Pat told Barry to stop acting like a kid, more or less.

The chatter around the fire slowly died and all eyes moved to the burly (or should that be tubby?) Barry, who was currently holding his chest and moving his mouth like a guppy. He was about forty-five years old, five feet eight tall, and fifteen stone in weight. Pat was three months away from seventeen, and three inches off his final height of six feet one inch. In stature he resembled a broomstick and he was forever pushing back his fringe because it covered his eyes. He also had a habit of bending his knuckles back until they clicked.

I don't think fear came in to Pat's vocabulary, but I was afraid for him. Pearcy was one of the men who stayed for the season working as a pole puller and, as it was a weekday, dad was back in London.

Pearce regained his voice and began moving towards Pat, swearing and waving his fist. He was used to hitting people, though usually he would pick women or children to vent his spite on. As his wife nervously stepped in front of him, I saw Tommy move to Pat's side, holding a piece of wood the size of a chair leg. This seemed to be a good idea, so I ran to the hut, came back with a rounders bat and joined them. We had a combined age of forty, a weight of about twenty-two stone, a couple of sizable sticks, and the fitness to keep out of his way, so I reckoned we had the edge. Pat took up a boxer's stance; I think he used to belong to a club for a while.

Eric Pearce began to cry and clutched his dad around the waist, telling him, between sobs, not to fight. Pearce looked at the boys in front of him, then at his burbling wife Doris, gave his young son a smack around the head, then turned and staggered off into the night, bellowing about the pain stopping him from giving all of us a good hiding. The sound of relieved laughter followed him and although that wasn't the last time he suffered from indigestion, it was the last time he made such a fuss about it.

Pat hated hop-picking, but his dislike of school was nearly as strong. He never did too much homework if he could get by without it, but I do remember he had to learn a poem once, to recite at school. It was 'Silver', by Walter de la Mare. It began, 'Slowly, silently now the moon, walks the night in her silver shoon.' The poet received the Order of Merit in 1953, which was possibly the reason for the interest in his work. Pat would have been fourteen then, so the time fits in. Anyway, Pat was walking around for weeks learning this poem. Anyone passing him in the opposite direction would have had an excellent example of the Doppler Effect. Like the change in pitch of a train whistle when it passes through a station, so Pat's poem would sound louder as he approached you ('this way and THAT SHE PEERS AND SEES silver fruit among silver trees') and quieter as he moved away.

He worked his socks off to get it right, but the peculiar outcome was that the rest of the family could recite it as well. We would not have put the Von Trapps out of work, but the sound of Walter could be heard for those few weeks throughout the house. Even Richard, who was only four, could tell us about the Harvest Mouse, scampering by, with silver claw and silver eye. The annoying thing is that while other events that may be relatively more important leave barely a mark on the memory, I can recall bloody 'Silver' word for word.

Talking about people suffering, there is a neurological disorder that seems to be increasingly common today. Symptoms include moodiness, anxiety, over-reaction in certain situations, shouting and swearing at people who have done nothing to antagonise them and a defiance to those in authority. It is called Tourette's syndrome, but I am sure that some people display these symptoms without having the illness.

From the age of about four to thirteen I lived in St Ann's Road, and just a tumbledown house and a few prefabs stood between us and Mrs Trett's. She was the lady who ran the super sweetshop that we visited every Thursday evening, It was dad's pay day and we were given a shilling to buy sweets between us. Usually we selected a variety of two ounce bags, but sometimes there were not enough sweets in two-ounces to give one each, which could cause a little friction, and Mrs Trett had been known to pop an extra sweet into the bag to make up the numbers. She would never hurry a child when selecting sweets, it was important to get the amount, the choice and the cost just right, and Mrs Trett's nature was diametrically different to that of Tourette's sufferers. She was quiet, helpful, polite and happy, and anyone seen displaying all of these symptoms should be classified as exhibiting Mrs Trett's syndrome.

Some mornings, after waking up and allowing time for the Fuzzy Felt to disentangle itself from my brain, one of my first thoughts was, what day is it? Until I was eleven and changed schools, if the answer was Thursday it would begin with a smile.

Thursday was games all afternoon at the rec, dad's boiled bacon dinner, followed by a trip to Mrs Trett's, and most exciting was another episode of *Journey into Space* on the radio. The point of this is to show that although nearly everyone likes weekends, Thursday for me was the only weekday that would induce a smile; whereas when we were 'Hopping Down in Kent', it was as if every day was a Thursday.

It has been many years since I've eaten a cobnut, and I'm not sure if they're still growing as prolifically in Kent now as they used to in the 1950s. It was then a common thing to go out along the hedgerows and woods and find an abundance of them. They were ripe to eat so they didn't last long once the hop-pickers had found them. They not only took them to eat immediately, but some of them went into a drawer at home to save for Christmas. I'm sure we also had a few apples and pears wrapped in paper and stored away until the festive season. It must have saved a little money.

At least the nuts are no longer used as ammunition for the catapults, but for one year the bushes were stripped and aggravation broke out. Pat and Tom were the prominent weapon-makers, carving the handle to the correct size for the person who was going to use it. I think they made them initially for target shooting, but the targets quickly became other catapult users. My brothers stopped making them, but poor copies were often found broken and discarded. The problem was that the user could easily hold the catapult incorrectly so that the ammunition hit the leading hand. Sometimes the elastic would not be extended enough and it would just flop out in front of the user; and most frightening was if it was pulled back too far and the catapult slipped from the hand and zipped back to smack the user in the face. All of the above could occur if the weapon was not made to fit the needs of the owner. Due to the amount of accidents that happened during the hop-pickers' 'year of the catapult', they were discarded by mutual agreement.

* * *

One Friday evening dad and Tom were driving down to the hop fields in an old Austin 7. They were making good time when the gearbox began grating. They tried to continue but were unable to select gears, then it would jam in gear. Fortunately, dad had a good knowledge of mechanics as he had been in the REME during the war. Most vehicle owners in the fifties had some experience of replacing parts, and even making them, if the problem was with a gasket or fan belt. Anyway, dad pulled up at a fish and chip shop, making Tom think they were stopping for an early lunch. Then dad went into the shop and asked the owner if he could spare him some flour. Tom's eyes performed a three-point turn as he pondered what they could possibly want flour for. Dad then poured it into the gearbox. When it had mixed with the oil that was left, it thickened up considerably and they were able to continue on their way. Tom said he never saw dad change the oil in that car, so if you're an Austin 7 owner, just check the gearbox for traces of McDougalls.

7

Abscess Makes the Pain Last Longer

In 1955, before I joined the others in Kent, I'd had a holiday in Middlesbrough with a mate from school. We went with his parents by coach to visit his relatives, and spent three weeks there, which was great fun. We met two girls, Jackie and Stella, and spent most of our time being shown around places that still trip off the tongue, like Redcar, Saltburn and Marske-by-the-sea. We also saw the cantilever toll bridge over the Tees, and of course the Yorkshire Moors. The first day we met the girls, they showed us the local park. I acted stupidly by trying to jump over one of the long seats that we passed. The very front part of my right shoe clipped the seat and the sole was torn clean off of it. I don't know if it was pride, embarrassment, or shame that made me do it, but I rolled onto the piece of shoe, scooped it up without them seeing, then, still acting stupidly, I dropped the sole into the bin. They were the only shoes I possessed and I had just cut their lifespan from several months to a week. I had no money for new ones, so I wrote home the same night pleading for footwear. I must be thankful to mum for responding so rapidly, and I realise that shoes were expensive, but any chance I had of a love life took a turn for the worse when I walked outside in those white laceless slippers which, as far as I knew, were only worn in the gym on games day.

Because of this holiday, I missed the journey to Kent on the removal van and travelled there by train the next day. When Christine saw me she asked about Middlesbrough. I told her about the girls. She went quiet then walked off and would have nothing more to do with me. She went back to school after two weeks and never came hop-picking again. I think it was a relief to us both.

The day after I arrived at the farm, I discovered that we had a man who would be calling two or three times a week from the shop in Petham. He drove an old brown Morris 1000 van with running boards along the bottom of the doors. In the back of the van he carried a few groceries, newspapers, paraffin and other essentials the pickers required. In the front, on the passenger seat, he had a pile of change and a few packets of cigarettes. Sometimes, when a little queue had built up and he was distracted at the back of the van, one of the youngsters would quietly open the front door and take just one or two packets of Woodbines. We would then go round to the back of the huts or across to the spinney and stroll along, puffing our cigarettes as if we were the bee's whiskers – or was that the cat's knees?

It became a habit for some of us to jump on the running boards just as he was pulling away. We could see he didn't appreciate it, so, like all youngsters, we did it more. But we always got off at the gate, so there were no problems. That was, until Ted and I refused to get off one day. We clung to the door handles and dared him to continue. Unfortunately he did, driving along the road, which was very narrow, with a thick hedgerow on either side. He must have been in a very bad mood, because he hadn't a thought about what might happen if we fell. He just drove the van at a furious speed, weaving from side to side so that we were whipped by the overhanging branches. He then slowed down and asked if we had had enough fun yet, and did we want to get off?

We were both cut and bruised, wet and afraid, but at that particular moment we hated that cowardly milkman. Then Ted amazed me by giving one of his all-time best answers, quietly replying that we would like the same again, but in

46

reverse. It was immediately apparent that the man's face was turning the colour of a fire engine. Realising we had gone a little too far, and not wishing to cause a heart attack, or get ourselves beaten to a jelly, we did the sensible thing and ran like whippets towards the huts. We looked back, but he was just standing in the middle of the road shouting after us. He must have considered this a downside of the job that was not acceptable, because a different man delivered the milk and groceries the next time, and his car did not have running boards. We never saw the first man again and the second one only called a few more times.

We never mentioned our little adventure, for fear of getting into trouble. However, two weeks later a large shed was put up near the manor house and each season it was there, selling newspapers, groceries, cigarettes and so on. The people who ran it were apparently from Kilby's stores in Canterbury. It was conveniently near and carried much more stock than the old Morris van ever could. So indirectly Ted and I were responsible for easing the workers' lives a little bit.

There was actually a more hairy moment to the escapade. On the way back we stupidly took a short cut across a large L-shaped field. There was one single tree in it, just by the angle of the L. We walked and talked about how good we felt now that it was over. I spluttered out his response to the delivery man, 'We'll have the same again, but in reverse.' We staggered about in tears of laughter and never saw or heard the bull trotting up until he was about twenty feet from us, then he bellowed. Without ever looking back, both of us ran for our lives. Clambering up a tree with a bull's horns a wafer away from your nether regions adds an unrepeatable strength to the muscles, but by the time the farmer arrived to take the animal in for the night, an hour had passed and those muscles were cold, stiff and itching from the nettles that had brushed them on the run for the tree.

Looking back on the incident with the grocer and his Morris

van, we had a very similar happening with another of the mobile tradesmen. Every Friday evening a fish and chip van visited the camp and, like the Morris, it had running boards, which seem to hold a strong fascination for young boys. Unfortunately, when Ted and I stood on them and refused to get off, we discovered he had probably been in this situation before. He just walked round to the back of the van, took a small bucket of tepid fat and threw it all over us. We slid off pretty quickly, feeling totally stupid. It was a difficult task to get the grease and the smell out of our clothes. In fact I think they were worn out before they became fat-free.

There was a little bit of 'own back' for us the following week, though. The chip man called as usual and, once he had his friers going on full heat and a queue of customers causing a great hubbub, we dodged around to the other side of the vehicle. Taking a wheel each and using matchsticks, we let the air out of his tyres. It was a slow job as we didn't want anyone to feel one side of the van getting lower. When we had finished we wedged a spud up his exhaust pipe for good luck then crept around the side of the end hut and watched. The queue seemed to get longer initially, but after an hour we were tense with anticipation as the last customer was served. The chip man took another five minutes to clear away his implements, then he got into the driver's seat and switched on. The engine burst into life, ran for a few seconds, then died. The driver got out, presumably to unhook the bonnet, but as he turned he noticed the flat front tyre. His cursing could be heard around the camp and five or six people walked over to see what the problem was. We joined them as if we had just returned to the camp. The chip man was telling the others that the engine had cut out and the offside front tyre was punctured.

'Offside back as well, mate.' The driver looked at the speaker, then moved towards him. 'You, it was you,' he shouted, a little hysterically.

Ted gave an excellent look of pained astonishment. He told

the man we had just got back from the village. Doris Pearce, Eric's mum, had seen us walking past the tap and backed up what Ted had said.

The driver appeared to be having trouble controlling his temper until Mrs Kerslake's son, Derek, handed him a foot pump to try. The man connected it to the front tyre and spent fifteen minutes turning his leg muscles to jelly. Derek offered to inflate the other tyre, while the chip man tried to start the engine. The van was parked so that the back faced the huts and as Chippie switched on and pressed the starter button, he blipped the accelerator savagely. The engine raced into life, the spud shot from the exhaust like a missile and crashed against Mrs Kerslake's door, with segments of shattered potato rebounding onto her head as she sat in her specially adapted wheelchair, taking pleasure in the evening's entertainment. She screamed for Derek to take her indoors. Derek, who had only put ten pounds of pressure in the tyre, ran across to her, leaving the pump attached to the wheel. The chip man could apparently take no more as he revved the engine dramatically and raced off into the night, with wheels spinning, flat tyre skidding, and the foot pump whipping in all directions until, after leaving several dents in the bodywork of the van, it snaked under the wheel and snapped off at the valve.

Ted and I just shook hands, aware that we had overwhelmingly won, but better than that was the fact that HE knew that we had won. Just as an additional sweetener we also informed everyone we met that the fat he used was rancid.

On the second Sunday, the four of us arranged to walk in to Canterbury. Like all good children we were wearing our best Sunday clothes, and as we were visiting the cathedral it was necessary to pass our mother's stringent cleanliness inspection. Ted was the smartest in his Prince of Wales check suit, and although it didn't blend with his black plimsolls, his idea that you never knew when it might be necessary to run fast was built on personal knowledge, and was much more important than fashion. We always took the same route to Canterbury. In fact it was a C road, more like a country lane; very few vehicles used it, and I don't think we passed through even one village before we reached Canterbury.

Most of our time was spent idly throwing sticks into trees, trying to knock down whatever fruit they grew. Or we would be looking out for any wildlife, different birds, grasshoppers, frogs or toads. For me that was the day's fun. Once we got to Canterbury, we seemed to do exactly the same things, like walk around the city wall, or go window-shopping without any possibility of buying anything as we had no money and it was Sunday. With all the deviations on the way, it took about three hours to get there, two hours to get back, and we spent about an hour actually in Canterbury. Sometimes if it was hot we took longer, so if we left the huts about 8.30 a.m. we would probably be back about 3 p.m., in time for a late lunch or an early dinner, and still with time left to get into trouble.

This particular day was enticing us to leave the road and sit under the trees. The heat was making us all feel uncomfortable, so we decided to get some apples from the orchard near Street End. Ted knew from previous visits that they were sweet and juicy. What he didn't know was that since last year the farmer had put up two fences to protect his crop. The outer one was just a couple of lines of ordinary cable. The inner one was three strands of barbed wire. We easily overcame the first by stepping onto a large log near the fence, then jumping over. The barbed wire was held up in the middle by one person, while the others crawled under.

It only took a few minutes to fill our pockets and get back under the barbed wire. The trouble occurred at the fence. With no log to climb on to, Ted tried to pull up the middle strand, making a gap to get through. None of us realised the fence was electrified and although it wasn't a very powerful shock, it was unexpected and made Ted stagger back onto the barbed wire. He didn't wait for us to help him, but wriggled into more of a tangle. It took us half an hour to stop laughing and ten minutes to free him. Ted was not so amused; he had a six-inch triangular rent in the back of his jacket and a six-inch triangular scratch in the skin on his back. The tear on his right trouser leg was big enough to display a lily-white kneecap, smeared with dirt. After we had finished imitating the laughing policeman, we piled some stones up to stand on and jumped the fence.

Back on the roadside, the decision was taken to keep going to Canterbury, where a needle and thread would be the first purchase. Strangely, although a shop was found that stocked them, they couldn't sell them to us on a Sunday. Fortunately, the shopkeeper thought, as we did, that the law was stupid. As Ted quietly told her in his near-to-tears voice that he would get a good hiding if he went home with his suit in that state, she took pity on his tears and not only gave him a needle and cotton, but helped him with the repairs. It took ages for them to finish but they seemed to make a good job of it.

The only trouble was, when we got back to the camp that night, as Ted bent to take off his best plimsolls, the knee of the trousers split open again due to the pressure. His mum was in the room and we all looked at her expecting our hair to be blown back straight by the strength of her outburst. She just said not to worry, the suit was second-hand when Ted got it and she would buy him some new clothes from the money they earned this season. Ted crept out of the hut with a red face. He was embarrassed at the offer and I knew that he would be vowing to work harder from now onward. I also knew that such a promise would last until he next got onto a hop field. Incidentally, when Ted began working at the age of fifteen, he was employed as an apprentice tailor. Today he still makes men's suits somewhere in Northants.

One of the things I always wanted to do on the farm was a

spell of polepulling, and for a very short time my wish was granted. There were three polepullers employed most of the time. Usually they came from the village and were full-time workers on the farm. Occasionally there would be a gypsy worker or someone from the huts, like Barry Pearce, doing it. This particular time we were working the field nearest the village and one of the gypsies was polepulling. Near lunchtime he was cutting some bits of bine down in our row and I asked if I could have a go. I successfully pulled down a piece of branch, then he said he had to pop off for a few minutes and I would be helping him out if I could take over until he returned. I was stunned, but seeing the potential glory in it for me, I agreed without hesitation. He left me the pole, which was about ten feet long with a sharp-edged hook on the top. I was ecstatic and swaggered around the rows where my mates were. They tried to pretend they didn't care that I was a polepuller, but I know they had imagined doing it, just as much as I had. However, it was becoming a problem now as he'd been gone for over an hour, and my nerves were being stretched like an elastic band, not least because I wasn't very good at pulling down the small pieces. Also, the pole was getting bigger. Like Pinocchio's nose after a bad session on the lie detector, so this pole was increasing in size and I was finding it difficult to keep in the air. I knew that if the farm manager saw me there would be trouble, especially for the gypsy. I was about to hide the pole, when he returned. He shook my hand, thanked me, gave me a shilling and made me dizzy breathing alcohol fumes into my face. The thanks, though, should have been to him, I never knew his name, but I think it was Jim, because although he didn't know it, he fixed it for me.

On the last but one weekend of this season Ted had us all sleepless for one night. He suffered from a toothache and it came on so bad that his left cheek and eyelid puffed up to a worrying size. Luckily for him it was on a Saturday, so the dads were all there with transport and Ted's dad was able to take him to the hospital at Canterbury. Ted had been building up to this for most of the day, first with a silent hangdog look, then with an occasional screech and finally, by tea time he was into full-blooded yelling, with the odd tear of self-pity. He wasn't even sure which tooth it was. It seemed to be all of them, conspiring to cause as much pain as possible. Ted rarely complained about anything, he just got on with it and spent a lot of his time laughing, so it was obvious that he was suffering.

I think Uncle Jim had an Austin A.35 at the time and Aunt Marge and my cousin Pauline went with them. It was like a family outing, but they should have taken a picnic with them as they didn't return until 11 p.m. Ted had an abscess under one of his teeth; that was what had made his face swell up. He was prescribed a course of antibiotics, and a pad was placed over his eye. This was held in position by a bandage tightly wrapped around his head. He looked very poorly and just kept holding his jaw, while the odd tear rolled down his right cheek and was soaked up by the bandage. He rather resembled the Invisible Man.

Most us were still awake when Ted dozed off. He slept for maybe fifteen minutes and when he woke up he had forgotten about the tooth. The bandages swathed around his head had fallen a little, and covered both his eyes. For a moment he thought he had gone blind and he ripped the bandage from his face. Perhaps due to the relief of being able to see, he began to sob. The outsiders went back to their own huts while Ted settled down and actually slept. I don't know what medication he had apart from the antibiotics, but he was back to normal by the morning to such a degree that he devised the hairiest game bar one that was ever played in Petham. He found a wooden ladder at the back of the huts. It was just a one-part, about twelve feet high. We carried it round to the front field, where Ted tried to balance it upright then climb up it. This was modified until two people, one each side, held it vertically. A third person then climbed it and, when the top was reached, the climber, with one leg either side of the ladder, had to balance there as the holders let go. Obviously the one balanced the longest was the winner, though marks could be awarded for grace on landing, just as time penalties were imposed for slow climbs.

The hairiest game we played required a sheath knife. Two players faced each other at a distance of two yards, initially with their feet as far apart as possible. The first person would throw the knife at the gap between his opponent's feet. The second person then moved whichever foot he chose so that it was touching the knife. They took it in turns throwing until one of them submitted or blood was drawn, in which case the injured contestant was the winner. Recommended footwear: extra large wellingtons with thick cardboard inserts.

Ted Hylott in the middle, his sister Pauline sitting, with mystery guests from the red huts.

A quartet of quality. From the left, Aunt Marge, Mum, Gypsy Rose and Hilda, who worked the next row to us on the hop fields.

Tally up time with, from the left, Mum, Mr Collard, a farm worker, Richard, farm manager and an unknown infant.

Uncle Jim training for the boat race with me in the front, then Ted and Richie astern.

This cheerful group are, from the left, Richie, Anita, Mum and Barby.

Our first holiday, at Canvey Island. Mum with Anita, Barby in front, then the white tide man wearing shorts that seem seamless.

Fishing with Pat and Ted near Latton Lock. Ted looks as if he's about to go swimming.

Young Ted with sister Pauline. Will they fill the basket before tally time?

Sadly, in October 2003, Mum died. Shortly before that she had said to Barby that there had never been a photograph taken of all of her children. She wanted one to be taken at her funeral, adding that we should all be smiling. From the left in descending age laugh with: Paddy, Tommy, Anita, Bobby, Eileen, Richard and Barby.

The bank of the River Stort, all the vegetation covered in hop bines.

The towpath by the Stort in 2003. On the left, buried in foliage, lie the remains of Latton Lock cottage and the big house, both destroyed by fire.

This cornfield was once the Banjo hop field, bisecting the huts and the spinney which, by the way, doesn't seem so aggressive here.

Second and third from the left, Uncle Jim and Aunt Marge, just returned from the village with a crowd from the red huts. The rest of the adults are probably still staggering along, one step forward two steps back.

8

All Scholars, No Ship

There was a problem for me the next year. Unfortunately, I had passed the eleven-plus and would be going to Tottenham County school. That news ended the chance of being allowed any extra holiday in the summer. It was necessary for me to be at the new school in time for the first register at the beginning of September. That allowed me just two weeks down in Kent. I was unhappy about it, arguing that the others had been given time off, but it made no difference.

I initially thought that the eleven-plus was an examination taken at sea, I associated it with a scholarship and I could see no other reason for the 'ship' part. At the time I wasn't even aware that I had taken the exam, it was only afterwards when we chatted about it that I realised. The pupils just entered the classroom one morning to find all the desks had been separated and papers laid out on them. I thought it was probably a class test and didn't worry too much about it. Perhaps that was the reason I had passed, I wasn't stressed until I found out later that the much talked about eleven-plus was over, with no sign of water anywhere.

There was, however, water during my first week at senior school, not tears, although I felt a constant longing to be with my family and friends in Kent. The water came, fortunately, during my first ever science lesson. Had it been any other class I don't think I could have continued to go there.

It had been usual at Stamford Hill Junior Mixed that if a toilet break was needed, a hand was raised and there would be no problem, providing it wasn't an excessive number of times. But I was warned by several people that senior school was different; students should be adult enough to wait until break times, and no teacher would let you out of their classroom during lessons. On the third day there I forgot to use the toilets before going in to lessons, the second of which was the science one. By the time this was halfway through I was desperate, legs crossed and praying I would hold out, but knowing that I couldn't. I had no idea what the teacher was saying, but luckily there was a large glass beaker in front of each of us, plus a tap and a Bunsen burner. Setting fire to the classroom seemed a little drastic, so I settled for filling the beaker with water while they were all doing some dumb experiment. Then, just as I cracked and began to pee, I pretended to trip and stagger back, tipping most of the water in my lap and what was left, over the boy next to me. The beaker slipped from my hand and smashed, by which time the teacher had reached my neighbour, slapped him around the head, called him a silly arse and told him to clear up the mess. I tried to explain that what had happened was my fault, but the teacher refused to listen, saying that the boy was a trouble-maker who deserved a week's detention. He clearly thought the lad, whose name was Mike Cook, had pushed me. It was not difficult to understand why. His philosophy was that if he smacked everyone around the head and called them silly arse, eventually all of the pupils would be punished for something they hadn't done and, more importantly, punished for something they had. That way, he considered, was fairer as everyone got punished.

I found it more difficult to understand the lack of protest from Mike, until I spoke to him the next day. After apologising to him again, he assured me that he really didn't mind and explained that at his previous schools he had been well-behaved all the time and knew that people thought of him as a bore. This silly arse business had given him a bit of notoriety, or street cred as it is called now.

My street cred was dismal to begin with. I suspect it reached its nadir when, following the smashed beaker incident, I was sent to the domestic science room with sodden trousers making me walk like a bow-legged cowboy. The search for the cookery room took a while due to the amount of teacher traffic along the corridors, most of them asking me what I was doing. I told them all that I was on my way to the domestic science room to dry my trousers in an oven. Most of them frowned and walked off muttering. I finally reached my destination and explained roughly what had happened. The teacher called me a moron and, to universal laughter, made me take my trousers off and iron them dry in front of the cookery class. It was just lucky that I had thought to put on underpants that morning. My mum had only just started buying them for me, because I was going to a 'posh school'.

I think my going to grammar school was a biggish event for my parents because they bought me a uniform, a satchel, sports clothing, maths equipment and other things that I might require. I can recall going with my mum to a schools outfitter in Bruce Grove and trying on the blazers and slacks, then picking out a real leather satchel. We went on to Woolworth in the high street and mum treated me to a pot of tea and a custard slice. It was very exciting, but the most unexpected occurrence of the eleven-plus saga was when Mr Baldwin, my form teacher at Stamford Hill, called me to stay behind after school and gave me a pair of cricket whites. He said that he didn't play much now and that I might have more use for them. I was so surprised I almost forgot to thank him. After a couple of alterations, notably to the waist, they fitted perfectly and I wore them for all the cricket I played during the rest of my schooldays. This amounted to a lot of games as I played regularly for both Tottenham County and Netteswell.

Apart from the headmaster, Mr W.W. Ashton, Mr Baldwin was the only teacher I remember from junior school. He was also the only one who took his class on trips, one of which was to his house in Woodford, where, after enjoying ourselves in the forest, we all went home with him for orange juice and biscuits.

There were actually eight of us from his class who passed the eleven-plus; five went to Tottenham County. One of these was another Arthur, everyone knew he was a certainty to pass. An indication of his quick thinking came when Mr Baldwin was setting us a test. He was writing the first question on the board and had just drawn a coin with the date 85 BC on it when Arthur was on his feet saying, 'Sir, sir, you've made a mistake. It couldn't be dated 85 BC. They wouldn't have known it was Before Christ because he won't be born for another eighty-five years, do you see, sir?'

Everyone groaned. It was frustrating, Arthur was always giving the answers before the rest of us had finished reading the question.

Of the three others who passed their eleven-plus from Mr Baldwin's class, one went to Tottenham Grammar and two to High Cross school for girls. About a quarter of his class passed, which seems an excellent rate to me. The fact that I passed, however, meant that I had to miss four weeks' hop-picking each year. Plus it cost my parents a tidy sum to kit me out.

Pat and Tom were both working then and only went down to the camp on a couple of weekends. They spent most of their spare time out with their mates in Tottenham. That first year at senior school I felt as if I had no friends in London, all of them were in Kent. Most of the time I felt abandoned, which could have been the reason for the trouble I seemed to get into.

I never liked fighting much, but sometimes during those early weeks I would suffer what could be called 'playground rage'. During the second week at Tottenham County the group of us from Stamford Hill were playing football in a corner of the playground, when a second year just picked up the ball and walked off with it. I was irate; who was this useless heap of dung? I ran after him and he turned, held the ball up and said if I wanted it I would have to come and get it. Without thinking, I kicked him in the nuts. As he doubled up and fell to the ground, he dropped the football, which I picked up. One of the boy's mates ran over and said that the boy on the floor would see me after school at the front gate. I then received a hard smack around the right ear and a plum-in-the-mouth voice barked, 'Silly arse.'

It was the bully-boy teacher, Mr Wood, aptly named as he was as thick as a plank. If he saw any boys fighting he would arrange for them to meet in the gym with boxing gloves on. He took my rapidly reddening ear between his finger and thumb and twisted until I was bent double. The pupils had all stopped what they were doing and were watching us. It was almost the end of lunchtime and Silly Arse Wood looked at his watch then smacked my head in time with his words, 'To – tomorrow – ow at one p.m. in the gym, and you too, boy.' He pointed to the lad on the floor, then strode off. Everyone turned toward the school and I took the opportunity, on behalf of underdogs everywhere, to give matey another good kick in the balls. He went down again, in tears. I pointed at him, touched my ear and growled, 'Tomorrow, one p.m. in the gym.'

I hadn't forgotten what his friend had said about being outside the front gates after school. I spent the afternoon worrying what would happen, and took my time gathering my books together. When I left, the area outside the school was almost deserted. There were just the two second years, some boys I didn't know, and a big Alsatian dog that my opponent's friend had. He pushed the dog forward so that it kept showing its teeth and barking. I was nearly in tears and the only way I could save myself from that embarrassment was to point at the one whose nuts I had kicked and say, 'I can't fight all of you and the dog, but I'll see you tomorrow in the gym.'

That evening, I told Pat all about my problem, and asked his advice. 'Keep looking at him,' he said. 'Failing that, hit him under the rib cage or punch him on the nose. If you draw blood or wind him, the fight's over.'

By lunch next day I was shaking. I got to the gym just as the others arrived. The teacher handed us a pair of gloves each. They were like pillows at the end of our arms and I couldn't imagine inflicting damage with them. Anyhow, I took guard, Wood rang a bell and everything Paddy told me evaporated. I ran at the other boy, who spun round and made for the door. I brought my pillow-clad arms down in frustration and the huge gloves shot off my hands. As he turned to see what was happening, one glove hit him full in the face, the other one dropped to the floor between his feet and tripped him. As he fell, his head hit the wall bars hard and he slumped in a sitting position, looking dazed. Wood wafted smelling salts under his nose, then decided to count him out. That was the only fight I'd ever had with gloves on, and I won it on a technical knock-out.

In the first year at Tottenham County we were given a choice whether we took woodwork or metalwork. Unfortunately, I chose the latter and spent three miserable terms with my ears being attacked by the screeching of metal on metal, collecting burn marks through my inability to handle hot objects safely and suffering a deepening depression whenever I looked at the thing I was supposed to be creating. Part of the problem was that I was never interested in whatever it was I was working on. In the last term at Tottenham County we were given a choice of several items. I elected to make a toothbrush rack, because it was small and looked easy to make. Wrong, it was small and annoyingly difficult to make. The dimensions on the drawing bore no resemblance to those of the rack. It was narrow where it should have been wide and only two of the four spaces would hold a normal-sized toothbrush. That was immaterial, though, as no one in the family owned a toothbrush, we just wet a finger, dipped it in salt and used that digit for cleaning. Needless to say, I switched to woodwork at the first opportunity. I wasn't any better at it, but I didn't have to suffer that appalling noise.

Noise was in abundance on the dog tracks, though, with the animals baying and squealing and fighting. The greyhounds could also raise a shindig at times. On a few occasions Pat took me racing, to Harringay, Romford, Walthamstow or White City. The track I enjoyed going to the most was Harringay. First, it was within easy walking distance of home. Second, we usually climbed over the wall to get in, so it cost us nothing. Third, two girls who were in my class at Stamford Hill junior school lived opposite the track and I sometimes saw them there. I think they were the girls who went on to High Cross school, Carol Horner and Christine Beezer.

To keep boredom at bay, while Pat was selecting the certainties, between races I would collect the losing betting tickets that the bookmaker had given to the punters before the last race. It appeared to be an unbreakable routine. Pat would take the money from his pocket, give it to the bookie, who then gave him a ticket, which a few minutes later Pat would throw away. This was actually happening all over the stadium and it was a pity I couldn't cut out the middle man, then they could just give ME the money; but that was a non-starter, like most of the dogs the punters backed. But what if he did lose his money? He didn't drink or smoke, and unlike those two vices, gambling would sometimes give him a useful return. Also, if Pat ever did have a winner, he always gave me a few bob.

It may be difficult to understand today, but some young children used to collect cigarette boxes, not just the cards inside, but the actual boxes, and the dog tracks were excellent places to build your stock up. If the cardboard back and front were cut out of the packet, they would be like second-class cigarette cards, worth perhaps ten for one real card. All collectors used to have a favourite, whether it was a dog, a soldier, sportsman or car. If their favourite was involved, they would never put that card up against any amount of packet-made cards. Well, perhaps they would do it once.

A boy at Stamford Hill juniors had been collecting cigarette packets at the dog tracks. He desperately wanted a card that one of the boys at school had, a footballer called Trevor Ford, who played for Swansea, Aston Villa, Sunderland and Wales. It was very rare in Tottenham, and could not be described as in A1 condition, but it *was* Trevor Ford. The boy had worked hard at three consecutive dog meetings and felt confident in making an offer. They arranged a game in the playground after school and, apart from the Trevor Ford card, all of the others were the cardboard cut-outs. Ten minutes later Trevor had changed hands. The previous owner was near to tears, but he had been unable to resist the amount of second-class cards that were offered, which was strange, considering they were never counted. But remember, the winner had been collecting for three meetings and when the first Ford owner asked how many the challenger would be offering against Trevor, the lad merely said, 'This many,' bringing a large shoe box from behind his back. As he took the top off, the neatly cut-out fronts of the Senior Service, Players, Woodbines and others sprang up, overflowing from the box. There must have been in excess of fifteen hundred cut-outs packed in there. Despite vowing never to use Trevor Ford in a game, the lad lost him shortly afterwards in a match that included John Charles and Billy Wright, but at least it was in a proper competition. This youngster would not have put up his favourite against cut-out cards even if he was offered a million of them.

It was easy to see the attraction in collecting or playing for these cigarette cards, though. They were colourful, educational and the matches were competitive. Sometimes, during break time or after school at Stamford Hill, there could be as many as thirty to forty boys either playing, or watching, in the small covered part of the playground. Providing you had a few cards to begin with, you had a chance to win up to a couple of hundred, which might well include several famous ones. Someone should arrange cigarette card matches for adults, and I think that meeting up for swap sessions (that's for *card* swapping), would quickly become very popular.

9

Fixed Wheel Fixed Me

Tom and his mates, mostly Brian and Colin, who lived in Eastbourne Crescent, used always to be out on their bikes until late evening. When Tom came in he would carry the bike upstairs and lean it against the passage wall. It was his pride and joy and he often paid for little baubles to go on it. Though pain is caused in the recollection, he once bought an expensive mirror for this keen machine, fitted it securely to the handlebars, then went out for the evening with Colin, who was temporarily without wheels. The fact that they caught a bus left me facing a big temptation – I know you can see what's going to happen, and I wish I had given some thought to it. Seizing the chance to have a ride on this super, blue, chromeflashing, drop-handlebarred BSA speedster, I carried it out to the road and, taking a deep breath to quell the nerves, I set off. It was the first time I had ridden it and I had no idea it had a fixed wheel, or even what a fixed wheel was. Apparently the cog on the back wheel continues to turn, even when the rider stops peddling. Well, I stopped peddling and was unceremoniously hurled over the handlebars, luckily falling inwards onto the pavement, I was helped up by a little old man, who then began inspecting the bike.

Luckily my wounds were minimal, but the shock was great, and it grew when I noticed the glass around me. I had broken the mirror and could already feel that bad luck encompassing me. There was no other damage and I carried it back because I was afraid to ride it again. I propped it up against the wall in the exact place it had been when Tommy had left it, then I called on Alan Lewis for a kick-about. I stayed out until just after dark, but Tom wasn't home until half an hour after me. I was in bed and feigning sleep when he did get in. Dad and Pat were also in, so I might get some protection if Tom noticed the breakage. How could I have imagined he wouldn't see it?

Almost as soon as Tom got upstairs, he picked up the bike to go out and a sliver of glass fell to the floor. He put the bike down again, looked at the mirror, and just kept on saying, 'That's my mirror, and it's busted.' He came in and pulled my indignant body out of bed and we all solemnly inspected the bike. Pat said it may have broken when Tom leaned it against the wall. Tom asked where the glass was then, and I cursed that I hadn't got the brains to bring the glass back with me to put by the wall. Tom grilled me for a while, but I never cracked. He said it had to be one of us, and I was the obvious choice. I yelled that it was probably one of Paddy's sidekicks out for a joyride. Pat went to smack me around the head and I quickly retracted that slur on his dodgy mates. The row went on for ages, and only now, writing this, can I admit to my major and embarrassing part in 'The Case of the Broken Bike Mirror'.

Dad was actually at home that night, but that wasn't often the case. He was a long-distance lorry driver, and he spent as many nights away as he did at home.

Usually, when I got back from school at about 4.15, I had a slice of bread and jam, then did my homework, or as little as I thought I could get away with. Pat and Tom returned from work at 6.30 to 7 p.m., and dad, as I mentioned, was sometimes away for two or three days at a time. One of us had to cook the meal, which may have consisted of burnt sausages, watery potatoes, an excess of peas and gravy that you could pick off of the plate. That was on the first day. The next evening, any sausages rescued from the previous meal were added to a bag of chips from Sino's and a slice of gravy, and so we went on, a gastronomic delight set before us daily. I had always had a ticket for free dinners at Stamford Hill, now, for some reason, I couldn't get one and I never fancied a school cooked meal as much as I did now.

The two weeks that I had spent hop-picking that year had been excellent. Most of the people there seemed to know I was going to Tottenham County and I strutted around like a peacock for a couple of days. Derek Kerslake brought me down to earth, though. He attended a grammar school in London and his mother, Doreen, expected him to go on to university, which was not a common event for a boy from a working-class family. Derek was a little older than most of the boys there, and only visited on the odd weekend. He spent most of his time studying, but no matter what his abilities were academically, he will be remembered by most of us as the best bow-and-arrow maker since those who supplied the English forces at Agincourt. He would use a five-foot length of wood, put a notch in each end, wind one end of a length of wire around the first notch then, pulling it so taut that the wood was bent nearly to breaking point, hook the other end of the wire over the second notch and twist it tight. The arrows he trimmed to a point, then weighted them and fitted them with cardboard flights. With a few of us in line, firing arrows down the field, it was actually easy to imagine we were English longbowmen fighting the French.

Derek handled his bow as if he had taken lessons, and if we had any competitions, whether for distance or accuracy, he always won. Then he began to show off about his ability and we stopped using the bows, and on the few times that Derek visited the camp, we tended to avoid him. We didn't need to do that for long, though, as this was to be his last season before he went on to university.

After the first week at my new school I was homesick for
Petham. I imagined what they were all doing, and they would
be doing it without me. That first Friday dad got home about
6 p.m. The arrangements had been for us to go down on the
second weekend, and that other week was stretched before me
like a month. Dad asked me what I would like for dinner, and
when I replied that I wasn't really hungry, he said what about
a juicy Kent apple, picked fresh from the tree. If we go now we
can be there before 9 p.m. I was washed and ready in less than
ten minutes. I conveniently forgot my homework. At 6.15 dad
was kick-starting the old Panther and by 6.30 we were listening
to the roar of the traffic in the Blackwall Tunnel.

At some point along the A2 there is a hill with a steep brow
that allows the traffic from London two lanes going uphill,
while vehicles travelling downhill have one lane. We were
chugging along, at a steady thirty miles an hour, and
approaching the top of the hill when, before we could see the
other side, a car appeared in front of us, overtaking on the
wrong side of the road. Luckily dad was sharp enough to
wrench the bike over to the left. As he did that, I started to fall
to my right, off the pillion seat. Dad was still alert and hooked
his right leg around my waist and thankfully stopped me from
falling in front of the oncoming car. The driver who had
caused this near miss accelerated off, leaving dad just shaking
a fist after him. We pulled off the road for a few minutes to
recover from the shock, then continued, and by 8.30 we were
climbing over the brow of another hill, this one leading down
to the camp site.

All the kids came running across the field to see who it was,
and as I slid off the pillion seat and joined them, I was as
happy now as I had been sad just two and a half hours before,
and thoughts of that near accident had almost disappeared. It
was strange, but each time I went to the camp for the
weekend, I never thought I was a part of the group until I had

been there a while, especially if anything came up requiring a decision of any sort. Arthur or Ted would take it, but I could see they felt a bit awkward too.

One thing I certainly never made a decision on was the surprise birthday party given one Saturday for Arthur by the red hut youngsters, with some help from the mums. They had laid a tableload of cake, biscuits, sausages, jelly, lemonade and so on. Arthur was twelve and had been hanging out a bit with the red hut crowd. Ted had been given the task of getting Arthur to the party by 4 p.m. without him finding out. As we wandered through the spinney about 3.30 the unsuspecting Arthur decided to pass the time with some tree-climbing. Unfortunately, the one he chose was part rotten. When the branch he was standing on snapped, Arthur, the record rope-assisted long jump holder, at that moment also became the vertical drop into a muddy ditch champion, with a fall of twelve feet. In the time it takes to blink, he was covered in a smell. He hauled himself upright, squelched out of the trench, and like a mobile compost heap, waddled away from us. A tear threatened to spill onto his cheek, but his dented dignity hung on by a shred. Ted ran after Arthur and obviously had to divulge details about the surprise party and persuade Arthur to go to the red huts to wash and get tidy.

The mums handled it perfectly. They sent Arthur into one of the huts, took his clothes, left him a bucket of water and some soap, then while they scrubbed the stench of rotted vegetables from his outfit, he covered himself in the hut occupant's eau-de-cologne. Once the huge fire outside had dried Arthur's threadbare clothing, he dressed and we all made our way to the cookhouse, where the food should have been.

The door had been left open and the mum who went in first gave a scream. One of the gypsies' lurchers was in the room. In fact he was lying on the table. The amount of food that had been prepared and laid out remained the same; the problem was that most of it was now inside the dog. He was chased all

68

around the camp, but the only edible things that remained were a few bags of crisps, some hard-boiled eggs, several sandwiches and the jelly. After Arthur had selected his plateful, we all helped ourselves to what was left, then sat around the fire singing.

Arthur, Ted and I had been going about together for so long now that we often thought of the same things as each other. In fact, there were times that we acted like brothers. It's funny, though, just how unalike some members of the same family can be regarding the things they enjoy. Tom had always found a great deal of pleasure in using any type of motorised transport, especially motorbikes and fire engines, which he drove for a lot of his working life. Pat, however, would just as soon walk, or catch a bus. He only really applied for a provisional driving licence because his mate, Ernie Chater, began pestering him. The pair of them had a window-cleaning round in Wood Green, and Ernie thought if Pat could drive, he could be responsible for the ladders.

Now Ernie had a really super black American Buick, which would be known today as a Babe Magnet. Pat was a little nervous about driving it, but matters came to a head one Saturday as the two of them were driving along Seven Sisters Road. Ernie saw a girl he had been keen on for several months. He pulled alongside her and asked if she wanted a lift. The girl said yes and got in the back seat. Ernie drove a short distance, then stopped and got in the back. He asked Pat to drive them just anywhere. Pat relented and pulled away. As he checked the rear-view mirror, the couple were completely unaware of his existence, but Pat couldn't take his eyes off them. He was a bit greener than Ernie when it came to 'adult relationships' and he perhaps thought he would learn a little from having the occasional glimpse in the mirror. But the hotter they got, the more Pat's concentration wandered and he ended up watching the back seat drama instead of the road and the beautiful Buick crashed, radiator first, into a lamp post. The couple were thrown in a heap on the floor and Ernie got up in a panic, asking Pat how he came to crash it and saying it wasn't insured for Pat to drive. The car's radiator was in a U shape around the lamp post. The girl, whose name Pat never knew, got out and made her own way home.

Ernie was permanently complaining about the money Pat owed him for the damage. When Pat moved to Harlow they sold the window-cleaning round and Ernie kept Pat's half, which he saw as paying back some of the car costs. Shortly afterwards Ernie moved to Hoddesdon and to my knowledge they have never met each other since.

Pat did not mind giving up the window cleaning, as it was not something he ever enjoyed, but I imagine he would have enjoyed the following event even less. It occurred whilst hop-picking, and required a vote from the family before it was included here. My apologies for the yucky content, but it did take place, though I'm not totally certain who did what and why.

There were five of us together on this particular Sunday. We were walking around the edge of the spinney and passing near to a tree that had forked branches. About once a week, when near the forked tree, one or other of us would clamber up it to a height of about fifteen feet, where there was a double branch formation sticking out from the trunk. By stepping onto it, dropping their trousers and squatting, anyone could make themselves comfortable enough to play the numbers game, and without nettles stinging all their crevices. The game had expanded since its conception and now the one in the tree could select one of the others to stand under him. The idea was for the one on the ground to look upwards, holding their nerve for as long as possible, then at the last second, leap out of the way. A bit like chicken is played on the roads. Usually the one on the ground could see when the bomb bay was opening and take evasive action. This day Arthur was quick to climb the tree and call Ray's name. In the time it took Ray to reach his position under the bomber, Arthur had stepped on to the fork, dropped his trousers and was squatting. Ray just had time to look up and say, 'Ready when you –' when a deluge that took the form of an open umbrella crackled from Arthur's backside at lightning speed, draping itself across Ray's head and shoulders like a cloak. Ray actually screamed and looked around, his startled eyes showing white and large against the deep tan he had suddenly acquired. He ran for the huts and water, while the rest of us were now rolling on the floor in hysterics. I don't think Ray ever forgave Arthur for that, because he must have been aware of what would happen. Anyway, the game was never played again, as no one would chance standing under the tree.

The embarrassment of that incident probably stayed with Ray for a while, but at least he knew he had the choice of whether to participate or not. The same option is not given to anyone who has ever been groped and however minor it may seem to others, they will know that the feeling of violation can last for a very long time.

Arthur, Ted and I were out on another of our Sunday trips to Canterbury and the heat was affecting us all. They were meandering along, under the trees on either side of the road, with their shirts wrapped around their heads like turbans. They were parallel with me but about twenty yards to my right. Each of them carried a piece of wood that was a cross between a walking stick and a cudgel, I was dragging my feet along in the middle of the road, and we hadn't seen a sign of life in half an hour, when I glanced up. Ahead of me, on collision course, was what I took at first sight to be a roly-poly man. About five feet four inches, fiftyish, almost bald and tubby all over. He grinned, but it came out a grimace. He was sweating profusely and stopped just in front of me, then, with a laugh like a braying donkey, he slapped me on the shoulder.

I was unsure what to do, but he seemed harmless so I laughed with him. Then, completely unexpectedly, he grabbed my testicles and made a loud guttural sound. Despite the heat, I went instantly cold and pushed him as hard as I could. As he staggered back his laughter became more demented. Ted and Arthur, seeing what had happened, ran towards me. I had begun to move away from the man and he hesitated, looked at the two youngsters waving their sticks, turned and walked quickly away. I shouted for them to leave him.

If I had known how many times that little incident was to affect me over the next few years, I might have opted for giving him a good smacking. At that moment, however, what had happened did not seem like a big deal, apart from making me feel a little unclean. I was in favour of forgetting it quickly and getting on to Canterbury. I tried to make Ted and Arthur promise not to tell anyone, but they insisted I should tell my parents. Ted eventually told them when we got back to the camp.

For the next three or four years, whenever a man came what I considered to be too close to me, I tensed up and prepared to hit him. A bus conductor once asked to see the Saturday football results, which were on the front page of the newspaper I had on my lap. As his hand came down to steady the corner of the page, I elbowed him and pushed him so that he fell backwards down the bus. He ordered me off, but I was already leaving. On another occasion I had spent Saturday afternoon playing football, then cycled to Leytonstone, where Ted and I went to the pictures. At some point in the film I went out to the gents. As I pushed the door open a sudden attack of cramp hit my right calf. An elderly man, who was just washing his hands, knelt down by me and went to push my toes back, which stretches the calf and eases the cramp. I punched him as hard as I could on the biceps nearest me, then ran out, leaving him gripping his muscle in pain. He was still holding his arm when he walked out of the cinema ten minutes later. Whenever I think of this time I feel guilty because I'm sure he was genuinely trying to help me, but the evil thought took over.

The event that helped me overcome this fear came when I worked on Soper's farm in Harlow. I had to go to Spitalfields market with Pete, one of the drivers. As he was unloading the lorry, a porter from the next stand kept grabbing at him and I wondered why the six-foot-four-inch driver put up with it, but finally he had taken enough. The next time the man went to grab him, Pete stopped working, turned slowly to the man and in a stentorian, town crier special he thundered, 'Look, if you want to hold my bollocks, then don't play games. Here you are, take them.' Everyone had stopped working and was watching as Pete dropped his trousers and Y-fronts and pulled up his shirt, so that his wedding tackle was on show to the world. Pete received a huge cheer, then he turned around and bowed, so that this time a hairy arse was mooning at the crowd. Another cheer, and the other guy turned scarlet and disappeared into the crowd as quickly as he could. I'm not sure why, but since that all happened, I've never again been bothered by those kind of antics.

Not long after Tom's mirror was mysteriously broken and everyone had returned from Kent, I had a very pleasant surprise. Initially, though, it was a shock. I had just got in from school. I opened the front door and bumped into a new bicycle. It wouldn't be for Pat or Tom, and from the size of it Richie was too young. A wave of excitement hit me. I heard mum walk across the landing upstairs and called her. When I asked her who the bike was for, she laughed and said, 'You.' Now I wasn't much into irony in those days and I didn't notice from the tone that mum was jesting, so I flew upstairs shouting that I was going to take it out immediately. When mum said that she was just joking, I was stunned. It actually belonged to Dennis the menace downstairs.

73

Mum was mortified that I had taken her seriously and a few weeks later, though I'm not certain there was a connection, both Anita and I had new bikes. Both of them were blue, with straight handlebars, three gears and a little saddlebag containing a multi-sized spanner and a puncture outfit. It was fantastic and I hardly let the saddle cool for the first week. Then I stupidly left it outside against the garden wall while I ran up to the lavatory. Two minutes later I was out again, ready to burn rubber, but the bike had gone. My desolation turned to anger and although I have never supported those in favour of capital punishment, for just a few minutes then I would have applied it to anyone even looking at my bike. Luckily for me, mum and dad had insured it, so within a week I owned my second new bike, which was perfect timing for the cycle ride to Petham.

10

Riverside Real Estate

Some time in April 1957 I had this great idea. A trip down to
Petham by bike, that would be quite an adventure. When I
told Ted and Arthur they seemed less than delighted, in fact
they fell about laughing. 'But it's miles away, we'd never be
able to cycle all that way.' About thirty-five miles, I had said,
knowing that if I had told them it was over sixty they would be
unlikely to come. They stopped laughing and asked how long it
would take. About three hours, I lied. I cajoled them into
coming by telling them how exciting it would be, seeing the
farm again, but without the hop fields. That was on the Friday
night, and we agreed to leave early the next day.

We oiled our bikes, packed some sandwiches and drink, told
a few people where we were going, then, armed with a map of
the south-east of England that I must have pinched from
Tom's saddlebag because he was the only one who never got
lost, we were ready to take the removal van route, and at
7 a.m. next day we pointed our wheels towards Kent and set
off.

The first ten miles went easily enough, but then the other
two began to complain because they had seen signs for towns
thirty or forty miles ahead, and began to worry that Canterbury was not mentioned, I told them to just enjoy the scenery
and if they got tired we would turn round and go home. After
twenty miles we stopped for a drink, which at least lightened
our saddle bags a little and helped us recover from the two
long hills that we had walked up. The rest of the journey was

arduous, with a lot of moaning from all of us, but after we had cycled over halfway it was easy to think that it was further to go back than it was to go forward, conveniently forgetting the return trip tomorrow. We did, fortunately, have a couple of long hills winding downwards where we could stick our feet out in front of us and enjoy the wind in our faces. There was also some indication that we were enjoying it more because, as we passed through Sittingbourne, we stopped for a while and actually considered spending a couple of hours in Sheerness, on the Isle of Sheppey. Then, realising how much extra legwork there would be, we continued to Canterbury.

We finally reached Chartham Hatch, which lifted our spirits because we knew how near we were to Petham, and then as we freewheeled through the village, we almost cried with relief. This was the only time we had taken our bikes to the farm, and although it took an eternity to get there, now that we had arrived, with bikes, we were able to get around much quicker. From the huts, which were already overgrown, to the village took half an hour to walk, but only a few minutes by bike. I was elated as I thought that for two boys of twelve and one of eleven we had really achieved something. Initially Ted and Arthur were happy and proud to have succeeded, but nightfall saw them a little less in the mood for singing.

We didn't have the nerve to ask if we could stay in one of the huts in case we were refused, so we cycled around until it was dark, then levered the lock from one at the bottom end, went in and sat on the wooden frame of the bed. We had no mattresses or blankets, so we left our clothes on and slept where we lay. I would like to have used Tom's bike for the trip; the tyres probably pumped up into air beds. Using our torches as little as possible to conserve the batteries, I was kept awake most of the night by the animal noises and the wind, and that was only from inside the hut. We ate the little bit of food we had left, to try to lift our spirits. As it began to get light we got on our bikes and headed towards the big gate.

There was a thick, grey, cold mist that was so low we couldn't see the wheels of our bikes. Fortunately, we didn't hit anything. We reached the gate where we had to begin walking up that first, long, steep hill that our kart had come down so quickly and disintegrated so painfully on. That was the first season, four years before this visit and it might have been happening now, it was so clear in the memory.

We could have cycled into Canterbury, but felt so tired that all we could think about was home, so we began our pedalling marathon straight away. The trip back as far as the Thames was very hard and, strangely, seemed to be all uphill again, the same as yesterday, but once we got north of the river, we all felt more lighthearted, until by the time we reached Leyton, tired as we were, we actually began singing. The hopping song, of course. When we wheeled into Edith Road it was late afternoon and some youngsters were playing in the street. One of them asked us where we had been. We looked at each other and we all began laughing. The nightmare was over, the dream was fulfilled. 'Kent,' we all shouted in unison.

In the spring of 1957 we moved from Tottenham to Harlow New Town. Except that Latton Lock Cottage, the house we bought, was hardly new. It was situated by the River Stort, between Burnt Mill lock and Harlow lock. The reason for the move was because dad was suffering from Hodgkin's disease, which affects the lymphatic system. The prognosis appeared to be one of progressive deterioration and it was probably felt that moving to the countryside would give us all a less stressful life. Dad may have seen how much we all benefited from going to Kent each year. If that was the reason, it certainly seemed to work, most of my spare time was spent in the boat, on my bike, or walking around the farm and sandpits.

I mixed in with the family much more than I had in London, although Ted often came over at weekends and holidays. Like me, he found the river and countryside gave him the same sort of freedom we enjoyed each year in Petham. It would be great to think that dad also benefited from the clean air and quiet surroundings. Unfortunately, I was not aware of how ill he was at that time. He was working in Key Glass-work, only half a mile or so from the cottage, and looked quite well.

The cottage was a little two-up-two-down that had about two hundred yards of river frontage, a large garden with half a dozen types of fruit trees, and a row of poplars towering over one end of the boundary between us and Pole Hole farm. The farmhouse was five or six fields away and the only access to the cottage was through the farmyard at the end of Redrick's Lane, Sawbridgeworth, and across the fields. Dad had to put railway sleepers across a few of the ditches to enable us to get the car and motorbikes out. A towpath ran the length of the property and it was a while before we got used to people walking along it, as if the garden was open to the public.

Although these day trippers and holidaymakers, many of whom remarked on how attractive the cottage was, would have looked at the advantageous things Latton Lock had, they may have overlooked the things it did not have. The four main, everyday things we had to live without were running water, electricity, a flush toilet cistern and a toilet with a roof on it. Next to the cottage were the remains of a much larger house that had been gutted by fire. Most of the walls remained standing but there had clearly been no roof for a few years, which accounted for the occasional sycamore growing through the foundations. Perhaps this may encourage some budding young architect to create a new form of tree house.

Strangely, a short row of wooden sheds remained intact and I wondered why these could not have been used for a toilet-bathroom complex, because the existing one was anything but complex. Still, at least it was on the ground floor, which saved us the job of building a ladder up to it, but it was roofless, doorless, seatless and flushless, thus ensuring it wasn't always odourless. Nevertheless, it was ideal for tanning the knees in summer, though a light dusting of snow in winter made the *Sunday Pictorial* feel like sandpaper. A kettleful of water was left overnight in cold weather so that the pump from the well could be primed to draw up the obnoxious-tasting liquid it contained. Most of the time a bucketful of river water was used for flushing the toilet.

Moving house to Harlow was, for most of us, an exciting time, even though changing schools was a pain. But for Pat it was all a pain. He was nineteen when we packed our suitcases, and all of his mates and the things he enjoyed would be left behind. Like greyhound racing, cinema, swimming, but mostly football. We lived midway between the Arsenal and Tottenham grounds and would go to watch the first team of each on alternating weeks. With huge crowds and no segregation, the games were usually very exciting. Pat played for a local team, and went fishing with his mates on the River Lea near Tottenham Hale. Plus he had other friends where he worked in Edmonton. He would find it difficult to commute the twenty or so miles each way. He accepted that Tottenham was not an ideal place for children, but he remembers trips to his local cinema, swimming pool, pie and mash shop, firm's parties, Christmas dances and a generally good time with his friends. He also, rightly I think, considered that Latton Lock was unfit for children to live in, with the lack of running water, electricity and proper sanitation. Initially he just refused to go with us and stayed at St Ann's Road. Then he had an argument with the Gumbritches, who obviously wanted Pat out, so they called the police. They got someone in Harlow to call and tell dad what was happening. Dad went straight to the old house and Pat was sitting on the steps, holding a box with some kittens in it. He returned with dad, but spent as much time as he could staying with his mates in Tottenham.

The first time I went to Latton Lock I was with dad. He needed to prepare the house for us to move into. He replaced some windows, fitted a pump to the well in the garden of the big house and rigged up a car headlight to a battery to give us some light. We were there for a week and the only things we ate during that time all came from tins, and were eaten cold: corned beef, pilchards and baked beans. I still enjoy beans, but the meat and fish courses have never been ordered since.

There was a small island with the cottage, plus a corrugated-tin boathouse that contained a long rowing boat which seated eight in comfort. It had clearly been a handsome-looking thing in its time, but its time had long passed. Much of it had a coating of some kind of moss on it. The oars were split in several places and if we'd had a rower amongst us, a few matchstick-sized splinters would have lessened the pleasure for them. Maybe they could have organised a new Olympic sport like boat baling.

We went out in it the first time we were all there together. After pushing it from the boathouse, we began rowing out towards the middle of the river. This coincided with the boat beginning to sink. There had been a small puddle in the boat when we clambered in to it. Tom, with all the insight of an early Kwik Fit fitter, suggested it had a slow puncture. All of those not rowing began baling out or stuffing handkerchiefs or socks into the widening gaps in the woodwork. Luckily, we were able to steer it alongside the steps on the bank outside the cottage before it was half full, as only two or three of us could swim. The boat was eventually pulled from the water and left in the garden until someone acquired the energy to repair it. From then on we all had lots of fun in it, rowing between Burnt Mill and Harlow to return lock keys, and getting friendly with the holidaymakers as they passed by on the river.

There was an elderly couple living in the house at Burnt Mill lock. They sold a small selection of sweets, so we tried to get into the routine that we had in Tottenham with Mrs Trett. Dad was paid on Fridays in Harlow and the weekly sweet ration had increased to two shillings and sixpence. The eldest three were all working so they no longer qualified, and their share was split between the four that were left. We would either row to Burnt Mill, or cycle. The choice that they had was so small, though, that we stopped going after a few weeks. Mum would bring home a few teeth-wrecking gob-stoppers or similar with the shopping, and the sweet-sharing sadly came to an end. At least, I think it did. Eileen used to eat her sweets so slowly that there is a possibility she may still have some of them left.

The walls of the big house did not remain standing for long. Ted and his family came to spend a weekend with us and, within hours of their arrival, a huge long rope that dad used to tie the loads down on his lorry appeared. One end was tied round the front wall, the other end transported across the river, along with all thirteen of us. Dad had been chipping away at the cement earlier and we proceeded to tug it down; in a very short time it was swaying, then a cloud of dust and a hearty cheer signalled the end of the first wall. The few dozen bricks that entered the water no doubt raised the bed along that stretch of the river, which probably explains why Pat never caught any fish where the big house casts its shadow.

The toppling of that first wall sorted out the women from the girls. Aunt Marge was an early casualty with rope burns. She went back to the cottage with Queenie (my mum) to make some tea. Luckily for them, they were inside before part of the second wall fell just short of the cottage. If they had been a couple of minutes later, they might have been left with a few bruises, and a long wait for assistance to reach us as the paramedics would have had to come on bicycles.

The bricks that had almost fallen on Aunt Marge didn't number many, but unfortunately they were all from one specific corner of the ruin and meant that the toilet, already short of so many vital parts, was now almost without walls. Before the crash, it was just possible to stand at the loo and see people walking by the riverside; after the crash, you could sit and watch. In fact, one man walked into the river through concentrating too hard on one of the girls sitting on the toilet. He scrambled up the bank with a helping pull from Pat, and stumbled off looking extremely embarrassed.

The other walls were eventually brought down, leaving a massive pile of yellow field bricks, half bricks, pieces of brick and brick dust. For a few months afterwards, all that seemed to fill our time was bricks. We had to chip all the cement off, leaving them clean. They were then sold for about five pounds per thousand. We all longed for the hop-picking season to begin. That never left you with dust-filled eyes and blood blisters on every finger. Still, brick-pulling was something I would not have missed, and for the two years we lived at Latton Lock, we had almost as much freedom and enjoyment as we had in Kent.

Latton Lock Cottage had been advertised in the *Exchange and Mart* for four hundred pounds. Dad agreed to pay ten shillings a week for sixteen years, a no-interest mortgage. It belonged to an insurance agent initially and when dad's health deteriorated, the man, seeing all the offspring, altruistically refused to take any more money. We were eventually moved into a council property near Potter Street, Harlow.

Moving to Harlow had entailed another change of schools. Compared to the school in Tottenham, Netteswell County secondary was a shambles. The sports section was good and the PE teachers seemed to enjoy their work, but most of the staff were unenthusiastic and had little basic ability. Certainly the pupils were not often prepared to work as a team with the teachers. It appeared that the students could not wait to leave, and the staff helped them as much as possible towards this end.

It was astounding to see the lengths to which the Youth Employment department would go to find a job that suited them when they left school. Each student had a maximum of twenty-five seconds spent on them. Each class of leavers was visited by two or three people from what is now known as the Job Centre. They pointed in rotation at each child, and asked what employment that person intended pursuing at the end of term. If the answer was something positive, indicating that they knew what they wanted to do, the employment staff would point to the table at the front and say that the leaflets there would contain any information needed. If the response was negative, they were told the leaflets might contain information that would help them. So, less than one minute of stardom, with no follow-ups, individual interviews or training offers. No personal involvement. It's a fairly sure thing that the employment staff did not know the names of more than one per cent of the children they were supposed to be helping.

11

Motorbikes, One Over the Edge and One in it

We'd been living at Latton Lock for a few months when Tom put his motorcycle up for sale. It was a Norton combination with single-seater Watsonia sidecar. Two young lads from Harlow came to look at it and it was easy to see that they would buy it. They just kept patting it and saying how great it was. I don't think they would have won much money playing poker. Anyway, most of Tom's possessions were kept in pristine condition, so I'm sure the bike was a bargain.

These likely lads paid the full asking price, then, with faces beaming, they jumped onto the bike and tried to ride it along the towpath. The guy must have misjudged the width, because the front wheel twisted down the riverbank and the rider refused to let go or jump off. The bike turned over as it fell and pulled the sidecar up into the air, until it was vertical. It stopped suddenly in that position and the youngster who was in the sidecar shot out like a champagne cork. He surfaced in the middle of the river and, thankfully, he could swim.

When they had both climbed from the water, their only concern was for the combination, of which we could just see the sidecar wheel, nestling just below the water line. Tom fetched a rope, which was looped around the wheel then thrown up the bank for us to heave on. Despite pulling until our veins almost popped, we hardly budged it, but luckily for us a couple of men were cycling along the towpath and they stopped to help. By pulling in unison, like a tug-of-war team, we eventually hauled it up the bank, where it stood, water draining away as if it were a colander. We thanked the men, who seemed really happy that they could help. Then the lads collected some old rags from the shed and took their time drying off the engine. After a couple of lusty efforts on the kick-start, the engine roared. They then decided it might be safer to push it along to the next lock and onto the road. As they shuffled away leaving a water trail a blind Brownie could follow, they continuously patted the bike as if it were a family pet. They deserved to get a lot of fun from it after their unusual method of testing that the fuel tank was waterproof.

In 1957 we substantially upgraded our Kentish cooking facilities. Our recent move to Latton Lock Cottage had meant a downgrade, as there was only Calor gas available, but at the huts in Kent there was no gas or electricity. There was a sniff from the older people when they saw us carry in the stove hooked up to a Calor gas bottle and begin making tea.

Mum had a run-in with one old biddy who was convinced that an open fire was quicker and better. Mum said she could cook the whole Sunday dinner at the same time, with up to three vegetables and the gravy on the rings and the meat in the oven. As it happened, she hardly had a chance to show her prowess as there was usually a queue of people at our door asking if they could just 'pop this little piece of meat in the oven' or 'just put this small dish of gravy on the top, it sometimes gets knocked over when it's on the open fire'. In the end almost everyone was coming to our hut with their Sunday roast. It's a wonder they didn't ask us to do their washing-up for them. Fortunately, the next year some of the others bought Calor gas stoves, so there was not so much pressure on mum (or her cooker).

It seems to have been a very eventful year. My birthday falls on the last day of August and we had been in Petham for a week by then. Usually it wasn't a big deal, just a card and some sweets or a book. In 1957, though, it was different; I became a teenager, and if I hadn't received the first present, I would have also missed out on the second, which would have left me a little morose.

Mum called me over at about 1.30 that afternoon and gave me ten shillings. It was for Ted and me to go to the cinema in Canterbury. If we hurried, we would catch the afternoon bus. I was elated, one minute I had to pick another three bushels of hops before I gained freedom, the next minute I was free. I ran across to Ted's lane and told him what was happening. He gave a whoop that for some stupid reason everyone began copying. We galloped back to the huts amid a cacophony of ye ha's and yoo hoo's. Five minutes later we were changed and on our way. The decision not to catch the bus was purely on biacomic grounds: that is, if we saved the bus fare we could afford to buy a comic.

The film was showing when we got there, so after spending ten minutes selecting an unhealthy excess of sweets, we went in. The title of the film eludes me now, but I think it was a western. Whatever it was, I thoroughly enjoyed the whole afternoon. As we left the cinema, Ted went into the gents and I walked slowly ahead, expecting him to catch up.

Coming out of the darkness into light made me blink rapidly. As my vision became focussed, I saw, on the opposite side of the road, Doody from the camp. She was with two other girls who, at that moment, went inside the shop. I took a chance that she might ignore me and said hello. I couldn't believe it. She turned around, smiled and said hi. We talked for a couple of minutes, then she asked what I was doing there. I explained why Ted and I were in Canterbury and she looked around, asking where Ted was. When she realised he was still in the cinema, she looked briefly into the shop, saw the girls were still occupied, then taking my hand she said, 'Quickly,' and pulled me round the corner. Her hands went gently to each side of my face and she drew me towards her. I was mesmerized. Those slightly parted lips possessed such a tenderness that after an embrace that grew in its intensity, I was left completely breathless. Doody then pushed me back, smiled again and said, 'I hope you have a happy birthday,' so quietly it was almost mimed. Then she was gone.

The whole episode could not have taken more than three or four minutes, and as I made my way back toward the cinema I looked into the shop. There was no sign of any of the girls. While Ted, standing on the opposite pavement, gazing blankly ahead, looked totally lost, which made two of us. I called him and he looked over, then crossed and said, 'Where have you been? I'm hungry.'

We had enough money left for fish and chips each, and Ted chatted on about the day and the film, then asked why I was so quiet. I thought of telling him, but with my pulse rate still into treble figures I decided that for the moment it would be my secret. Besides, I'm not sure that Ted would have believed me.

One of the more disliked characters on the hop fields was under pressure; we never knew why, just that it was to do with a motorbike. Barry Pearce occasionally rode as a pillion passenger on a blue and cream 250 Norton Jubilee. Nobody knew who the bike was owned by, but he was thought to be a drinking friend of Barry's from the village. Barry was out riding with his mate one Saturday evening, heading back to Petham, when the rider missed a bend. They skidded, spun round and crashed into a huge privet hedge. Barry was wedged in the bushes and had to be helped by some pedestrians, who pulled him free from the other side. The motorcyclists were unhurt, apart from minor cuts and bruises. When they eventually got back to the huts, Tom saw the bike and asked them how the damage had been caused, and whether it was the same bike that had been involved in an accident in Petham earlier.

Apparently the macho Barry turned a 'whiter shade of pale' and pleaded with Tom not to let his wife Doris discover he had been riding a motorcycle. As far as I know, she never found out and, on the assumption that it's no longer important, I will just clarify things. Doris, if you have been wondering these many long years why Barry came home that day with a bird's nest in his pocket and looking like the Scarecrow in *The Wizard of Oz*, this was the reason: he was quite literally pulled through a hedge backwards.

I was out so much of the time with Ted and Arthur that I didn't see much of the rest of the family except when we were actually picking hops, or eating. On one occasion, though, Richie, Barbie and some of their friends had been scrumping near the village. They were carrying their bags of ill-gotten gains back to the huts when a local man saw them and shouted that he was going to call the police and have the thieving bastards nicked. Everyone began to run for home.

Just inside the big gate and a little to the right there was a duck pond. Barbara was lagging behind the rest when in a desperate attempt to catch up she climbed the fence in the corner of the field, then jumped, straight into the duck pond. The ducks flew at her, quacking and squawking. She was apparently the only one without any scrumped apples, which was fortunate. She half swam, half waded to the far side, but before she could get out, a policeman had gripped her by the arms and marched her to the police car. They drove to the huts, where everyone, including the youngsters, was around the fire, singing. Barbie was told to remain in the car. She waited for the policeman to reach the huts, then she took off her drenched jumper and wrung it out on the driver's seat, so that both front ones were sodden. The policeman spoke to some of the others, then looked in a few of the huts before leaving Barbie and driving off to Petham. I could almost hear the squelching as the clothes on his backside sponged up a pint of stagnant water.

One bag of the apples had gone up the cookhouse chimney, the other in the unused mattress in the top hut. The police had not found them, but it seemed they only wanted to give a warning that they would be keeping an eye on us in the future. That was the only time I ever saw the police in Petham, but of course scrumping a large bag of apples must rank very highly on their list of crimes to warrant turning out. I can appreciate the annoyance the farmer feels if the trees are damaged, but on the whole we were very careful scrumpers.

Talking of damage being done to trees makes me think of the only time that I attempted to increase our stock of firewood. Along one side of the garden at Latton Lock Cottage, there was a row of poplars. They had been cultivated either as a wind break for the house and cottage, or as a part of the boundary with Pole Hole farm. We'd been living there over a year, and that particular day I was looking for something to do. We always needed wood for the fire, so I decided to chop down a tree. There was one that seemed a little too close to those on each side of it, so to extend the gap and give the outside ones a chance of some growing space, I decided to take the middle one out.

The chopping at first was easy; the axe was sharp and heavy and sizable chunks were flying out with each hack. I was alone in the house that day as everyone was either working, at school or shopping. I estimated I would have at least two hours to myself. However, after half an hour I was exhausted. I had a rest, but couldn't stop for too long in case the tree decided to fall at an inconvenient time. I had tied the gates up at each end of the towpath, and left notes advising people using the path to proceed with caution. Luckily, no one came by in all the time I was working there. I took out a couple of deep, wedge-shaped chunks, put the narrow end into the gap and hammered it further in, using the reverse side of the axe.

I was beginning to get very nervous and wished profoundly that I had never started. My arms were shaking with the effort and every other swing of the axe was like hitting a piece of solid rubber. I rested again for ten minutes then decided to give it everything. I checked both ways of the towpath; nothing was in sight, so I returned to the tree and began chopping with all my strength. A few minutes later, as I was about to sag again, I heard a slight creaking noise. I quickly looked along the towpath again; nobody about. Then I jumped the ditch and ran along beside the other trees. I knew if it fell this way I would have some protection. I stopped about five trees along and looked back, just as the creaking increased. Then in an awesome, slow-motion display, the tree began falling, in just the right position – except I had misjudged the length and the top twelve foot or so was suspended over the river. The tree was being supported by the thick branches around its middle, which enabled me to stand on the trunk and chop off all of the smaller branches. From being almost unable to move my arms, I was now in a frenzy to get the tree off the towpath before I got into real trouble.

I had cleared a reasonable amount when I suddenly thought of Pat's old car, I think it was a Humber. It was a big black motor, untaxed but a good runner. Most of the time he only used it in the field behind the cottage, like a racing car on a track. The keys were always in it, and the first turn of the ignition had the engine purring. I backed it up to the tree and attached one end of a rope I found in the sheds to the back bumper, then the other end to the tree. I pulled slowly away until the rope was taut. The car shot forward and the bumper shot backwards as they parted company. I retied the rope to the back axle. I think their shape made the branches act like skis on the hard surface, because the tree actually began to move. Not far, and the bottom end slid into the ditch, but it was almost clear of the towpath. I carried on until the family began drifting in, then I returned the axe to the shed and waited until I had recovered until I tackled it again. I'm not sure if anyone noticed, but it was several days before I cleared it up and stacked the wood by the cottage.

Also, there was no mention of the car being used, even though I had only balanced the back bumper in place. I think Pat didn't have the car on the road because of the cost, plus Tom used to take him about, which on one occasion had an unfortunate outcome, especially for Pat, who had recently been selected to play in goal for Harlow Town.

That particular evening, when they were returning home from work in Edmonton, riding on Tom's 350 BSA Gold Star, they were involved in an accident. They were going over some crossroads, when a van came across from the right of them and hit the side of the bike. Pat's right leg was badly broken by the impact. A helpful householder called for an ambulance and stayed with Pat until Tommy had got to the hospital on his bike. After Tom had found out what condition Paddy was in, he rode home to Latton Lock, arriving about 10 p.m. to an understandably worried family.

The van driver was taken to court and found guilty of driving without due care. Paddy was compensated for several months off work and pain suffered. The leg had to be broken and reset after a few weeks as it was not repairing properly. He also had almost a year out from football and, considering how much he loved the game, he should have been given a substantial pay-out for that alone. It was also fortunate that the girls he met from then on did not all want him to race them. The limp he was left with after the accident would have ensured that he retired a virgin.

With a little of the money Pat received, he bought a Pye radio, which was operated by a 9V accumulator, re-charged weekly in the electrical and bicycle shop in Old Harlow. It was great to be able to listen to our favourite radio programmes again. The most popular time to tune in was Sunday afternoon. Our meal was finished by 1.30 p.m. and the radio was placed in the middle of the table. Everyone flopped down on their usual chairs, the floor, or stairs and concentrated for the next two hours on *Ray's a Laugh*, *The Goon Show*, *Round the Horn* with the Glums, and a real weird one, Peter Brough's *Educating Archie*. How on earth can a ventriloquist have a show on the radio? Yet it was very popular.

Most of us were listening in, when there was an unusual scratching sound at the French doors. As Tom opened them, mum fell through, completely exhausted and unable to speak. She had just returned from her friend Stella's in Edmonton. Walking along the towpath, a bag of shopping in each hand, she saw and heard someone following her and she ran home. While the rest of us chased around seeing that mum was comfortable and making tea, Tommy picked up the air rifle that we kept downstairs and ran out into the darkness, firing it everywhere. The amount of wildlife that must have taken a pellet that night probably went into double figures.

Happily, mum recovered very quickly. We notified the police, who spent a little time on the case, and we discovered later that the tracker was the local farmer, taking a short cut across the fields. Clearly no action could be taken as he was merely crossing his own fields to get home, but every time mum visited Stella after that someone would be at the station to meet her.

It was hard on mum, the amount of work she did, but dad did try to ease it a little. We had been at Latton Lock for about a year when dad bought an old cast-iron boiler. He fitted it in the out-buildings and it was used from then on to heat water for washing our clothes, or taking a bath. Dad had so much energy even when he was sick, that if he had been well when we lived at the cottage I think he would have renovated the big house. With Pat, Tom and Richard's (even though he was only eleven) abilities to fix anything mechanical, electrical or just needing plain common sense, plus my undoubted skill at labouring, we could have had the best lock house on the river. As it was, Tom took a job at Key Glass-Work where dad was working, for the princely sum of three pounds, seven shillings and sixpence, but, as he said, he didn't have any travelling expenses.

12

Power Mad with a Yellow Hatband

In 1958 three things happened nationally that affected me to some degree. The first left me stunned and dismayed for a long time, as I'm sure it did most people when they heard the news. It was the Munich air disaster, which occurred in February of that year and involved Manchester United's football team. They had just played a cup game in Munich and were leaving for Manchester when their plane crashed on take-off. I had just returned home when Pat told me of the tragedy. I found it difficult to accept the magnitude of it, especially from a sporting point of view. Eight players were killed, including three English internationals, Roger Byrne, Tommy Taylor and Duncan Edwards. There were also back-room staff and sports reporters among the dead and injured. There was an air of disbelief and gloom everywhere. For weeks the newspapers were overflowing with details of who had died, who was injured and whether they would play again. It was the one thing outside family matters that had a profound effect on me.

The other two events were related to each other. On the sixteenth of June 1958 the first yellow lines were painted on British streets. The colour was clearly not chosen for its aesthetic value, as a stream of custard along the gutters could hardly be anyone's first choice. Colour apart, the reason for the lines was discovered less than a month later, when money-eating machines called parking meters were introduced to Londoners. They seemed to just grow up from the pavements overnight. I don't recall there being much discussion about it;

one day you could park without let or hindrance, the next a bunch of grim-looking individuals were patrolling the streets issuing tickets. That initially sounded exciting, but imagine telling a friend about it.

You: 'I was given a parking ticket today by one of those new wardens.'
Friend: 'So what does that entitle you to?'
You: 'To park somewhere I suppose, but you only get one if you're on a yellow line.'
Friend: 'What yellow line were you on then?'
You: 'The one on the warden's hat.'
Friend: 'So someone knock it off his head?'
You: 'They did. I parked on it and won this ticket.'

The English motorist should have insisted they could be planted on our pavements with the proviso that only foreign money could be used. This would have dispensed with all those coins with harps on them that the banks refused to accept and ensured that foreign drivers would at last be paying something towards the upkeep of the roads that they drive so poorly on.

Pat passed his driving test before the nation's kerbstones were so delicately and colourfully enhanced, while every couple of years another of the family earned the right to add to the chronically overcrowded streets of Britain. Anita, whose spirit of adventure led her to take the test in an army truck, was once passing near to Harlow and got permission to leave the convoy to visit mum. There were a few surprised people around Red Lion Crescent when she almost filled the road with this huge lorry. It was a challenge just to climb up into the cab. I know she enjoyed her time in the Army, though she may have joined up initially to escape the housework and cooking she was expected to help with as the oldest girl in the family. It was quite unfair really, as I never saw Pat doing much bed-making or dusting, which was a pity as he would have looked cute in a pinny.

Pat's dislike of hop-picking was internationally known, but there are two things that he does enjoy. Number one has always been a little flutter, followed closely by the pleasure of collecting. From toby jugs, to plates, Royal Doulton china or clocks, if he thinks there's a bargain to be had, he becomes a proper little Lovejoy. He usually makes a fair profit as he travels around the markets and auctions, but his favourite place for earning money is at car boot sales.

For all of the people, like Pat, who get up at 5 a.m. to grumble in the jumble, here are a few lines of a poem I have written just for you.

The Sunday Supplement

Each Sunday at the crack of dawn, with shuffling gait and stifled yawn,
citizens seek their Holy Grail, at the local, vocal, car boot sale.
The sellers, vehicles packed with tat, like fish hooks and unworn cravat,
must, on arrival, overcome a refereeless rugby scrum.
In the soundless tick of a timeless clock, the seller loses half his stock,
before, unaided, he repels this horde of squelching infidels.
The burger van, just lighting up, sells waxy tea in a plastic cup.
While rancid fat begins to drape, across the scene like a giant cape.

The seller takes on buyer mode, purchases an old commode,
several books and a string of beads, a box of plants overcome by weeds.
A childhood game, three pieces short, a cracked decanter for special port.
A bunch of artificial flowers and a bent umbrella to evade the showers.
Much, much later, around midday, when it's time for the seller to pack away,
a tiny problem comes to light, he's got more now than he had last night.
Still off he drives with his new antiques, more to sell in the coming weeks
and the acting aphrodisiac? He'll soon be buying his own tat back.

Eileen was the second member of the family to join up. She met Dave, her husband, while serving in the Royal Air Force. They travelled extensively, and as Dave's work included arran-

96

ging sporting events, barbecues and summer camps, we never really knew if they were working or on holiday. Eileen is very sports-minded and has played for the county at hockey and netball. She is also very considerate, and if information is needed about someone in the family, she is the one to contact as she keeps in touch with everyone. Like Anita, Eileen saw hop-picking as thoroughly enjoyable and has many happy memories of those times.

Richard, on the other hand, never seemed too keen on it, perhaps because he was only four the first season, so would not be expected to have many memories of the early years. The boy nearest to his age group was seven-year-old Ray Lewis. On the few occasions I was with him, I was hardly the best of brothers, cursing him and calling him distasteful names, for which I now profoundly apologise, though most people encounter aggravation from their siblings, and I received as much as any, but I'll take the regrets as read.

He did, however, recall a phrase carved into the wall of Latton Lock Cottage: 'Men may come and men may go, but I'll go on for ever.' We both assumed this was referring to the river, as the cottage was totally destroyed by fire during the year that we moved out.

13

The Harmonica Player and the Grey Lady

This chapter briefly outlines a few of the recollections that dominate some of the Lawson family's memory banks. Most of them were remembered by some of the people. Some were remembered by most of the people, but the one that was remembered by all of the people, I'll tell you about later.

Every year, the trip to Petham went without a hitch, in the furniture van packed with all our belongings and the families singing at the top of their voices. For some of those regular hop-pickers, and include Tom, Anita, Eileen and Richie in that group, this may be one of their most vivid memories. There was some question as to whether such a journey would be allowed to take place today. I don't know about all those people packed in the back of a removal lorry, without seat belts: they might be arrested as an unsafe load, plus, if their voices are no better now than they were then, they would almost certainly be bound over to keep the peace.

On the topic of memories, Anita mentions the earwigs that crawled up your legs when you went to bed and being warned that they were inclined to find an ear and burrow their way through it into the brain. There are some things that just do not improve through being remembered, and earwigs definitely come into that category. She does, though, have another memory that strikes a chord in many people. That is of getting up early in the morning and going over to the hop fields when there was a mist in the air ('The Grey Lady'). All the bines were so wet, whoever pulled them got a soaking, and unless it turned out to be a sunny day, they remained damp and uncomfortable for hours.

Many people still vividly remember Tom for the audacious manner in which he acted when seeking solitude. Perhaps he knew then that in fifty years' time people would be asked about their memories. Well, the one that earned Tom enduring fame was when he built a chair out of bales of straw, sat in the middle of the cornfield and taught himself to play the harmonica. What would he have done if he had been keen on the piano? Anyway, the general feeling from those who had views on the matter was that they were unsure whether the chair or Tom was the Chippendale. It's difficult to believe that no one in the talented group who occupied the black huts ever went on to win *Stars in their Eyes*, although Tom apparently came very close to winning the Turner prize for art with his old corn chair.

Three or four times a season opportunity knocked for some of the kids when, as Anita, Barby and Eileen recall, they put on shows in the cookhouse. They insisted on having a curtain separating the stage from the audience, in spite of this decreasing the available seating space by up to forty per cent.

A lot of work and many good ideas must have gone into the staging of these productions, but unfortunately any knowledge of them slipped from the short-term memory bank some time in the fifties. However, I have been advised that when audience figures dropped below six, it could have been because all the potential money-paying public were backstage waiting to participate in the show. It was felt by many – well, some – OK by Anita – that her unusual voice might lead her to the stage and stardom. Clearly the talent scouts never travelled as far south as Petham.

For Anita, and Eileen, thoughts of the village shop were enough to whisk them back fifty years in a flash. They recall clearly the times they went with mum to buy provisions from the little store that had such a wonderful smell about it. Anita likens it to Aladdin's cave in the eyes of young children, with all the sweets, groceries and comics. A further delve into the smell memory convinced them both it was a mixture of fruit and veg and paraffin. I think they are both in need of some aroma therapy.

The fifth of November bonfires weren't the only ones that stirred the blood each year. Almost everyone was strongly attracted to the fires that were built each evening, first for cooking on. These were made so that they had just the right sized logs to be able to rest the pots and pans on, as there might be up to eighteen of them cooking at any one time. It wasn't unusual for one of the pots of food to tilt over and spill its contents into the burning wood. That meant baked potatoes and butter again. Then after the meal, chairs were brought out to encircle the fire and usually a sing-song began that could last the whole evening.

One of the events mentioned by several people, including Ted and Anita, I have no recollection of at all. Patients from the mental hospital near Petham were allowed to work in the fields in the summer, and this of course encompassed the six weeks of hop-picking. The wardens stayed with them each day and apparently brought large packs of sandwiches with them for lunch. No doubt there was some variety, but those who mentioned the episode said that sandwiches left at the end of the day were given to any worker who wanted them, but they only remember thick, cheese ones.

The reason not everyone recalled this may be because we had picked our quota of hops around mid-afternoon. By the time the 'sandwich spread' was taking place, we would have been enjoying our freedom. The times we did not have much of that was when it rained heavily. It was back to the huts for everyone and we had to dry off as best we could, usually hanging our wet clothes around inside the cookhouse. The big loss, I think, was not being able to sit around the fire and chat, so some of the pickers would pop into next door, leaving half of the huts crowded when it rained and the other half empty. That was not a common recollection, so perhaps most people only remember the sunny days.

When Pat was asked what his best memories of hop-picking were, his answer was brief, 'Going home.' He did recall, however, as everyone did, Friday evenings around the fire, waiting to see dad's headlights approaching from Chartham.

If these memories were in competition for the most humorous, I suspect the following, contributed by Tom, would be amongst the favourites to win.

Just as we kept a lookout for dad's car, so it must have been for all of the others gazing up towards the hill. Most of the dads tried to make it for the weekends. A couple of times we had fraught waits when someone's father didn't arrive. Once, on a rare occasion that Arthur's dad had written to say he would be visiting, his car refused to start and he couldn't get it fixed until the next morning, so he arrived a day late. The other time one of the men from the bottom huts had a puncture and didn't have a spare or a jack. Fortunately, he managed to repair it at the roadside. He found some bricks to hold it up, while he dug a hole under the tyre to get the wheel off. Luckily someone who stopped to help, as many people did in those days, had a puncture outfit and tyre levers with them. Although the inner tube was competently repaired, it took him far longer than usual to get to the hop fields, so a few people were worried for an hour or two.

There seemed no way of letting people know, until some clever clogs suggested giving everyone the telephone number of the Petham taxi service. If anyone was stranded after that, for any reason, they would call the number and the cab driver would make his way to the huts with a message, collect his fare and everyone was happy. One of the men decided to test the system and outlined his idea to the others. They went through the details on the Sunday and the following Saturday lunch time they all drifted about the camp and began disappearing. Amazingly, not one of them was noticed as, one at a time, they made their way to meet the others in the spinney. When they were finally all together, they walked to the pub in Petham. The person arranging this test phoned the number for the taxi driver and asked him to get a message to the wives at the black huts that their husbands were in the village, getting drunk and playing around with the local women.

When the driver delivered this news to the wives, they were thunderstruck. Until then they all thought that their husbands were with them on the camp, or in bed having a lie-in. The thought that they could be getting drunk and messing around on a Saturday lunch time, while their wives were just a couple of miles away, was beyond belief. They had one last look for them. Someone even tried calling, but there was no sign. The women were now getting angrier by the minute and they all squeezed into the taxi and headed for a showdown, threatening all kinds of vengeance on both the husbands and the village women. When they arrived they all made for the pub. Of course when they went in the men were quietly sitting at a table, sipping orange juice with not a woman in sight. They all kept straight faces as they asked their wives what was wrong. They couldn't keep that up, though. In light of the ladies' grim looks they explained that they had been conducting an emergency test, and everyone had passed.

The men then fell about laughing and after a few drinks most of the women joined them. Anyway, the emergency contact call worked perfectly, though strangely, I don't think they ever used it again.

One memory of the biggest pillock at the hop fields was implanted near the end of the second year. I'm talking about Barry Pearce, of course, and his almost as clueless son Eric. It was late Sunday afternoon, Ted, Arthur and I had returned from Canterbury and as we approached the camp we heard shouting coming from the middle area of the huts. It looked as though there was a fight taking place. We ran to the cookhouse and found Eric swinging a lump hammer at the wall. His tear-stained face made it seem as though he had been there some while and the blows were so weak he was hardly marking the stone. His dad renewed the taunts he had been making, snatched the hammer from him and began pounding the stone and chanting about a thing worth doing is worth doing well. As bits of it began chipping away, Big Bazza lashed the wall with all his might and several pieces were dislodged.

l asked one of the onlookers what was happening. Apparently Eric's dad had given him a whack earlier and in a fit of temper Eric took the hammer and began hitting the wall, though not hard enough to damage it. Now there was a hole Big Daddy could step through, all down to the moron Barry. He suddenly realised that a lot of people were watching him and he had the good grace to leave, looking a little shame-faced, but he never repaired the wall and within two years few of the bricks were still standing, and it became an eyesore.

14

The Language Teacher

The last week in August had two important dates in it. First, it was always the start of the hop-picking season. Second, it was my birthday on the thirty-first. We began our yearly trips to Kent in 1953. Uncle Jim and Aunt Marge had told my parents about it, how much could be earned and the freedom for the children (which, of course, also meant freedom for the parents). Mum and dad were sold on the idea, though as you will see, it was very fortunate for me that they did not speak to Jim and Marge the year before.

On the thirtieth of August 1952 my dad took me for a birthday treat, to one of the greatest football stadiums that has ever existed, I mean, of course, the marble halls of Highbury. Arsenal were playing Sunderland and dad was a firm Gunners fan. The size of the ground overawed me and when this group of red and white clad men ran out to a deafening ovation from sixty thousand spectators, the hair on my neck prickled up. From that moment The Arsenal were my team through thick and thin, or in some seasons, thin and thin. There was such a crush during the game that, along with many other youngsters, I was passed over the heads of the fans in front and allowed to sit on the grass surrounding the pitch.

The day was not a total success, because Sunderland won 2–1. But Arsenal did win the league title that year. Dad was raving about Jack Kelsey, Cliff Holton and Alex Forbes, but the player who seemed to be the darling of the crowd was Sunderland's Len Shackleton.

That season and several to come, I stood outside the stadium midweek and collected the players' signatures. It is now fifty-one years since I first saw The Arsenal and I have enjoyed watching a host of great players, from Logie and Lishman, George Eastham and Joe Baker, to those double-winning teams of 1970/71, 1997/98, and 2001/2002. The last contained so many world-class players that usually those sitting on the bench would have been sure of a first-team place with any other premier side. The Arsenal will continue to be my team as long as they and I exist. But I wonder if I would have followed them had we started our hop-picking adventure the same year as the Hylotts. One thing is certain, I would have missed out on years of pleasure.

As far as I can recollect, Tom never supported any professional football team, though he did enjoy playing. Apart from his wife Lyn, the thing he loved most was driving. He learnt to drive in a green 1953 Consul, but took his first driving test in a Ford A type station wagon, which dad had converted. As the examiner climbed in, he whacked his skull on the back of the bulkhead and spent the first ten minutes moaning about a lump, 'the size of an onion' that had altered the contours of his head, leaving it 'like the face of the moon'. Clearly that did not include the Sea of Tranquillity. Needless to say, Tom failed that test. He passed at Bishop's Stortford in 1959 in a driving school's Morris Minor and it was the second happiest day of his life.

In May 1960, aged twenty-one, Tom got a job with a Cheshunt firm, driving a sand and ballast tipper lorry. He was there until he joined the Fire Service in November 1961. Dad died on Tuesday the sixth of September 1960, the very day that Tom got a new job erecting fences. When he got to the yard he had a phone call all about dad, and left straight away for the hop fields. Earlier that year he had been driving 'green goddesses' and Commer pipe carriers with the Auxiliary Fire Service. His brigade number 60549 stayed with him until his retirement in October 1989.

We moved to Red Lion Crescent in early 1959, after dad had approached the local council to rehouse us. To take all of the furniture and other household effects from Latton Lock, dad didn't do the conventional and hire a removal van, he borrowed a horse and cart from someone he knew locally. With that leading the way, followed by dad, mum and Barbie in the Wyvern, Tom and Pat on their motorbikes and the rest of us on our bikes with whatever we could carry, I bet the residents of Red Lion Crescent uttered a communal 'Oh, no' as we clip-clopped up to unload what belongings we had at number twelve.

We all found it difficult to accept immediately that we had gas, electricity, running water and a flush toilet. We kept drinking from the taps and switching the lights on and off. We also had more space, though all of the excitement these things engendered were more than counteracted by the fact that dad was unable to get around much. He had a room downstairs and mum gave him round-the-clock care.

Red Lion Crescent seemed to be deserted during the week. There was a small grassed area a little way up from us, where the local children played football after school. I'm not sure how they managed before we moved in, as the Lawsons usually formed one of the teams. However, there were no small knots of adults outside their homes, chatting about the day's events, as there had been in Tottenham. No groups of youths riding their bikes in slow, lazy circles while they thought of places to go. No couples walking slowly around the crescent, holding hands and occasionally stopping for a flattery-charging kiss. In fact there appeared to be no heart in Red Lion Crescent in comparison to St Ann's Road.

Weekends were slightly better, but only because several of the men in the crescent would be out lying under their cars. I'm not sure if they ever did anything except cover their clothes in oil, but I suppose just jacking up the motor and lying beneath it on a cushion would enable them to keep out

106

of their wives' way for a couple of hours, or get their heads down for a sleep.

A few weeks ago I had reason to call near the crescent, so I detoured round to see if it had changed. The only thing that appeared different was the lack of noise. No barking from Dandy, the little black cross-breed dog that we got from the gypsies in Kent so many years ago. No sound of the large privet hedge springing back into place after I had failed to retain Olympic gold in the decathlon. Mind you, I had a good excuse; there was a four-inch-deep gutter exactly where my take-off point for the high jump was. Finally, there was no sound of children arguing coming from the house. Just the silent silhouette of a memory brought to life after half a century of waiting.

Maybe it was the pleasure I derived from working on the farm in Kent that made me want to take it up full-time. At seventeen I applied for a job at Soper's farm in Harlow and was taken on as the jack of all trades, but none of them skilled. Mostly my work was in the big barn, loading and unloading lorries with the foreman Charlie Dellar and a highly strung Cornishman called Vic.

The first day that I was there, at 1 p.m. everyone took their sandwiches (or Cornish pasties) into what seemed like a giant greenhouse. I was sitting alone, not sure what to do, when one of the field workers, a brute of a man called Fred, began speaking loudly and animatedly to some of the others. He was liberally peppering his sentences with the strongest swear words and there were at least a dozen women there. My dad and his generation had always abhorred bad language in mixed company. 'Ladies present,' he would admonish. Usually the person swearing would apologise to the women and use a little less florid language. This seemed like such a situation. Shaking a little, I went up to Fred and quietly asked him to stop swearing as there were ladies present. Everyone stopped talking as I began, so I actually chastised him in this silent room, with two dozen pairs of eyes burning into my back.

Fred certainly stopped. His mouth dropped open and he gazed at me, not comprehending what had been said. The silence grew and my face was on fire, then Fred began laughing and laughing. Unable to drag in enough air between each boisterous gurgle, he almost choked before someone whacked his back and got him breathing normally. I didn't think what I had said was funny and told him so. Still spluttering, he could only point at the women and croak, 'Silly bastard.' Then he told me that next to them he was Mother F****** Teresa. The ladies were all looking highly embarrassed, but were not in the same league as me.

He was right. In only a couple of days I heard and saw the women field workers say and do things that made my hair curl. Their favourite form of amusement was to jump on any young lad who strayed too near to where they were working and strip his clothes off. Soper's staff engendered the type of environment that made growing up fun, and after one week I had acclimatised and thought they were all gorgeous, though completely unpredictable. There was no dare, truth, or kiss in the barn where I worked now, but it was showing an awful lot of promise.

A few weeks after my introduction to comedy, I was almost saying hello to tragedy. Two young lads worked in the fields bringing in the crops that the women picked. This particular day they were collecting boxes of rhubarb. One of the boys drove the tractor, the other one stacked the boxes on the trailer. The driver was supposed to stop the tractor at each little group of boxes, walk round and lift them up for the other boy to position, then get back onto the tractor and drive to the next stack. Of course nobody did this. The driver would jump down while the tractor was moving along in the deep ruts that were always there, pass the boxes up for the other lad to pack and only get back to driving when the tractor needed turning at the end of the row.

Luckily for the driver, on this occasion it had been raining recently and the earth was spongy. As he went to jump down at the beginning of a line, he slipped on the foot rest and fell lengthwise into one of the ruts. His foot was immediately trapped under the huge back wheel and the tractor began to slowly roll along the length of his body, squashing him into the mud. He screamed for help and his partner jumped down and ran round to see what had happened. He was clearly stunned by what he saw but kept remarkably calm. He climbed up into the driving seat, stopped the tractor just before it reached the shoulders then reversed it back until the lad on the ground was freed. The boy now doing the driving had been yelling for help all the while, and one of the pickers ran to phone an ambulance. They had to literally peel him from the earth. His head was swollen and almost blue in colour, but with just one week's treatment in hospital he recovered fully and returned to work at Soper's a couple of weeks later, though it's a good bet that he will, in the future, always stop the tractor before he gets off.

I had been working on the farm for about one year when I did a pretty stupid thing. I got on well with all of the people working there except one of the drivers. He was a bit flash, always talking about how good he was at just about everything, especially driving. When he parked up at night he was the first in and always took the back space on the left looking in to the barn. There were four trucks that fitted in snugly, two at the back and the other two at the front. Initially I was full of admiration for his driving. He always accelerated rapidly as he backed into his space and, using just his mirrors, he parked in the same spot night after night, always stopping inches from the back of the barn. I mentioned this to Charlie Dellar, the foreman, who just laughed and pointed up at the roof. A length of string hung down, with a tennis ball attached to the end of it. The reason it was there hit me suddenly. He just reversed his lorry up until the ball touched his offside mirror and he knew he was less than a foot from the back of the wooden barn. He plummeted in my estimation and I felt stupid listening to him talk about his skills.

A couple of nights after pointing out the tennis ball stunt, Charlie had to leave early. Whenever he did that he left me to lock up the office. After he had gone, I arranged some of the sacks of potatoes into a form of steps, as the barn must have been twenty feet high. These enabled me to reach the ceiling without too much difficulty. I had to be there early next day to unlock the office, but fortunately the drivers had all left for the markets by the time I got there. Before Charlie arrived I was able to climb up the sacks, undo the string that the tennis ball was attached to, then retie it about eighteen inches nearer the back of the barn. I placed the sacks of potatoes back in their previous position, then spent the day worrying about what I had done.

The intention had been to alter the distance enough to have the truck just touch the barn. Well, I could only have been a couple of inches out. At 4 p.m. that day, after he had loaded his truck for market, he jumped into the driver's seat, started the engine, blipped the accelerator and roared back into his parking spot. Once again he stopped exactly as the ball touched his mirror, but this time his tailgate went through the barn wall and Vic, working outside, was showered with the boxes he was stacking. People came running to check what had happened. Fortunately, nobody was hurt and the old boy who mended the market boxes managed to re-assemble the jigsaw of wood that was the back of the barn.

There was an enquiry into what had occurred, but I just kept quiet. Charlie was the only one who seemed to have guessed what had happened. He wagged a finger at me and told me to be careful or someone might get hurt. Stupid as it seems, I still sometimes wonder what the driver must have thought when it happened and how Cornish Vic must have stepped back in amazement as the back of the lorry suddenly appeared, and the neatly stacked boxes all flew into the air.

15

An Attempted Wage Snatch

One other unusual event happened while I was still at Soper's farm. It was around Friday lunchtime. One of the drivers, Len Webber, was on the back of his lorry, about to begin stacking several tons of King Edward potatoes which I would send up to him on the loader. This contraption had a never-ending belt that the sacks were fed onto then carried up to the driver, who packed them into place. The lorry was not quite under the loader and Len asked me to back it up a little. I got into the cab and started it up, and at that exact moment a scruffy middle-aged man ran from the offices, jumped on one of the bicycles leaning against the barn, and peddled off furiously. Len yelled down for me to follow him, and I saw in the rear-view mirror six or seven people hurrying out and shouting that the man had stolen the wages.

I followed the cyclist as well as I could on the deeply potholed farm road. He stayed in front up to the main road. Lenny was bouncing up and down like a rubber ball and banging on the roof as if we were the posse, yelling, 'Faster. Catch him!' If the man had turned onto the main road, it would have been dodgy to follow him, because although I drove the lorries around the farm, loaded them and parked them, I wasn't old enough for an HGV licence. He obviously didn't know that and he cycled across the road and into the recreation ground, his sense of panic urging him on. That was quite flat, so I easily caught him now.

I was just keeping level, awaiting instructions, when Len

pulled one of the skeins of rope down from its hook. For a moment I thought he was going to unravel all the coils, put a loop in the end and lasso the thief. Nothing so time-consuming, he just whacked the guy across the side of the head with the thick rope, knocked him from his bike, jumped over the side onto him, then held him until some assistance came. That didn't take long; half of Soper's workforce steamed in, carried by a variety of vehicles, including two tractors and a 1935 army truck. The man made no attempt to run, but Len said afterwards that he wished he had, so we could continue the chase. It was an extremely exciting few minutes in my life as a farm hand. We were all given the thanks of 'the guvnor' (tug forelock) but I think any of us would have paid to take part in what was undoubtedly a successor to the Keystone Cops.

My time at Soper's was the second job I'd had since leaving school at sixteen when dad died. I was first employed in a factory in Harlow, then after a few months of smelling like an oil can, I went to work on the farm in Old Harlow. Almost every minute I was there I enjoyed, and for all of those eighteen months I was going out with my first real girlfriend, Elaine. Her father, who was a director of a company in Harlow, was intimidating, as were the large house they lived in and the gymkhanas she competed in. I knew I would need a better job if I was to continue seeing her, so I took a course in selling. That same year I passed my test for both car and motorbike and shortly afterwards bought a Bedford Dormobile twelve-seater. It had a top speed of about 40 m.p.h. and was always causing problems. Every Saturday about eight of us used to go to a dance hall, usually the Tottenham Royal, where the Dave Clark Five used to play, but after the acquisition of the Dormobile we were free.

One of the venues we visited was in Nottingham, a dance hall with a rotating floor. If one of the dancers caught your eye, a short wait would bring them past you again. We had an excellent evening and a fish and chip supper, before leaving about midnight on the long trip back to Harlow. Unfortunately, it began to rain and my wipers were so pathetic I was barely able to see. Consequently I didn't notice the deep pothole until we thumped down it, then bounced back from the kerb onto the road. A deafening rattling noise indicated that the exhaust had fallen off. I pulled up, put on a raincoat and slid under the van. My brain was clearly not in gear that night, because without thinking I gripped the exhaust to fit it back into the front part. I yelled in pain as my fingers sizzled on the hot pipe and grabbed the underpart of the door to pull myself out. Unhappily it was a sliding door and one of the lads pulled it open to see what I was shouting about. The door slammed against my fingers, cutting two of them quite badly. Not knowing this, I wiped the rain from my eyes. When they pulled me out, hands and face covered in blood and oil, they all gave cries of concern. One of them was so worried about me, he asked how they would get home if I had to stay in the hospital.

After I'd tidied myself up and waited for the exhaust pipe to cool, I got under again and used some cardboard to wedge the two parts together, then bound it with a kiddies' snake belt that one of the blokes was actually wearing. We all piled back into the van and tentatively drove off. There were no problems until we reached Sawbridgeworth, four or five miles from Harlow, where we free-wheeled to a halt with an empty petrol tank. We were near a garage, but it was closed. A couple of the group decided to walk and apparently got home at about 6.30. The rest of us tried to get some sleep in the van before the garage opened at 7 a.m. We finally got to Harlow town centre at 7.30. We had a football match that morning with an 11 a.m. kick-off, against one of the top sides in the league, Greaves and Thomas. Amazingly, considering none of us had slept for more than an hour, we won 5–0.

* * *

That Dormobile gave me two of my most embarrassing moments while driving. The first was after I had pulled up to give a lift to a hitch-hiker. He was going to Potter Street in Old Harlow and asked me to drop him at the shops. It wasn't too far, so I assumed he was in a rush. As we pulled up by the shops, he slid the door open and jumped out. The door didn't close properly, so he tried again. The third time, he was determined to shut it. He drew the door back so hard that unfortunately it slid off the runners, landing with a thud on his toes. He lifted his head and gave a wide-eyed look of pain. I jumped out, ran around to pick the heavy door up from the pavement and put it in the back of the van, as he bent double and just hopped around moaning, like a contestant in the dwarfs' race at Butlins. A lot of people were now watching and I asked him if he wanted to get back in the van for a while. He looked at me, still with that lack of understanding, or perhaps it was shock. Anyway, I took from his silence that he could cope. As I got back in the van I called out to tell him that there was a doctor's surgery above the grocer's if he needed one. I drove off and saw him in the rear-view mirror, limping slowly to a bench by the bus stop. After waiting about ten minutes I drove back past the shops again, but I couldn't see him anywhere. So if you think that you are the hitch-hiker, my profound apologies are for you. Unless you want to claim off my insurance.

The second time I felt stupid was while driving to Harlow from Alders Green, a village in the outbacks of Hertfordshire where the air is thick with bewitching blossom and the inhabitants are mainly just thick. When I started out it was very cold, with a mist that cut visibility to a few yards. I drove slowly and twenty minutes later I had reached Bishop's Stortford. I had my headlights on as I was still finding it difficult to see. I could hear horns being sounded behind me, but I had no chance of accelerating. In fact it was now like a London pea-soup fog. I was only driving at ten miles an hour, so I decided

115

to pull over and wait for the weather to improve. I carefully parked and slid open the door as a queue of traffic began accelerating past me. It was a crisp bright day, with unrestricted vision. The frost on the outside of the van windows was now almost dried off, but as there was no heater in the vehicle it had been a slow job. The inside of the windows had been totally steamed up then frozen over, so I could hardly see through them. Several of the drivers in the tailback that I had caused gave me a blast as they passed. All I can say is that nobody called me anything worse than I called myself, and whenever I drive in fog now, I always wind the window down a little.

Shortly after I left Soper's farm I started a doorstep delivery service selling pre-packed potatoes around Epping and Woodford. A guy called Steve, who I knew from football, helped me out initially, but it really didn't need two of us, especially since as soon as we earned a few pounds we went to the snooker hall in Leyton and played until we were broke again. After a couple of months I really required someone to assist me financially. The camel's back was finally shattered by a road tax renewal request.

At that time I hadn't paid mum any keep for a few weeks and had actually come to blows with Pat, who, quite rightly, thought I wasn't pulling my weight. Mum rarely got on to me about money, which probably made me even lazier. I think most of the family took mum a bit for granted, but I was the worst. Unfortunately, I only began to realise this when I left home to work at the holiday camp in 1966. I saw for perhaps the first time the work she did for me in terms of washing, ironing, cooking and so on. Multiply that by the number of people at home, plus she had a full-time factory job to hold down.

I think she also must have dabbled in homespun psychology,

because I recall an occasion when I took a shilling from her purse. The next day, as we walked to the hop field, mum was behind me, talking to Aunt Marge about the missing money. Marge asked who she thought might have taken it and mum replied quite loudly that she only knew that it couldn't have been Bob, as she trusted him completely. I didn't hear any more, but I felt so guilty, the first chance I had I returned the money to the purse, and never, ever took anything from her again. It was many years later that I read Freud's psychoanalytical theory, which focuses on guilt or moral anxiety on one side, and pride or self-esteem on the other. I don't think mum ever studied it, but I'm sure she knew how I hated feeling guilty, and the thought that she trusted me prompted me to return the money. I'm only sorry that I didn't have the courage to give it to her personally, with an apology.

I imagine the feelings she would have had about what I did next, just to raise a little money. I decided that the only way I could stay solvent was to 'borrow' a few sacks of stock from the farm. I could perhaps pay it back anonymously when I'd raised the money. There were always eight tons of potatoes ready for market each day, and Steve was surprisingly happy to help for thirty per cent of the profits. The plan was for Steve to wait for me at the end of the drive, in the Dormobile, then to follow me when I drove past in the truck. We would head for Steve's house in Stratford, unload most of the stock into the garden, drive to Woodford, offload what was left into the van then, leaving the lorry, both take the van to Harlow.

One morning I left the van, with Steve in it, about 3 a.m., walked quickly to the offices and found the keys to the truck, got nervously into it and pressed the ignition. The noise startled me and I stalled the engine, which took another two turnovers before it burst into life. As I rumbled along the drive I saw twin beams of light searching the sky, like a clip from the Battle of Britain. I reached the road and turned towards the lights but my curiosity was quickly doused. The back end

117

of the Dormobile was nestled down in a ditch by the roadside, and the front end was pointing at Pluto. I later discovered that Steve had become bored with waiting for me and just for practice had decided to execute a turn in the road. He didn't see the ditch he was backing into until the rear doors hit the bottom of it. I dragged some rope from the truck, tied one end to the front of the van and the other to the lorry. Within fifteen minutes we had extricated it, and Steve was wittering on about whether you could see the headlights from the moon.

We drove to Stratford without any further mishaps. Despite the fact that Steve had not yet passed his test, his driving was quite good. When we arrived at his house he told me that I had driven all the way with my headlights on main beam, so he hadn't been able to see in his mirror. With just a pinch of sarcasm I asked how he knew it was me, if he couldn't see. His eyes glazed over. I saw that he was searching for some snappy response, but I didn't have that long. We had barely loosened the ropes on the tarpaulin, when a slow, measured footfall made me shiver. Unmistakably size ten, belonging to a bobby on the beat. Who else would be walking the streets at this hour of the morning?

I hopped up on the back of the truck and carried on pulling back the cover. It seemed that to continue as if we were genuine delivery men was the only way out of this predicament. Then, just as the stress was getting to me, I realised the sound of the footfalls were getting quieter. The policeman had turned off and was walking away from us. I almost whooped with joy. That made my mind up and I began to retie the ropes. Steve asked what I was doing and I explained that we had been so fortunate this far, the van in the ditch, the headlights on main beam all the way, the policeman turning down a side road and so not seeing us. Any of those could have landed us in court and perhaps in a juvenile detention centre. I realised now what I could lose by carrying on with this stupidity.

I finished tying off the rope, then got into the driver's seat of the truck and told Steve to follow me. I only drove a few miles, then parked the truck in a side road and took over the driving from Steve in the Dormobile. I drove him home, and although he earned nothing from the night, he seemed quite happy. He told me that he had really enjoyed himself and if any other jobs came along, just to give him a bell. I thought of the number of times I had suffered an irregular heart beat that night and I marked Steve down as a definite fruitcake.

After I'd made up my mind not to continue this lazy life, a feeling of elation gripped me. It lasted right until I began working at Greenpar Engineering the following Monday.

16

A Life Revisited

In 1957, with Ted and Arthur, pushing our bikes up the hill towards Chartham at the very start of our trip back from Petham to Leyton, I remember a few tears blurring my eyes. This was where our kart had crashed four years before, where I had left two lines of skid marks, decorated with the skin from my elbows. Those four years had passed as if they were four weeks. Now it is fifty years exactly since the time of the accident and I can see the faces of those kart pushers, who had seemed so intent on generating enough speed to create a sonic boom. That was the twenty-ninth of August 1953. Now it is the twenty-ninth of August 2003.

My son Ren, who is thirty-one, reminds me that there is no Dorian Gray type painting to keep me from ageing, and the nine-year-old boy racer is now a fifty-nine-year-old dreamer with a head full of memories that he needs to recall. Ren had, as a birthday present, arranged to take me on a two-day trip around these memories. Our initial target, two and a half hours from Bournemouth, was 90 St Ann's Road, Tottenham, North London, where various members of my family had lived between 1946 and 1957.

As we left the congested M25 to join the traffic playing honk the horn on the A10, two things were noticeable. The first was the vast increase in the amount of vehicles, which for those living in London must mean attaching the car to a tailback every time it is used. That is, of course, assuming you can find somewhere to park it between journeys. I have a vision of some poor souls just driving around for twenty-four hours a day, searching for a space big enough to take their Mini. The second fact relates to the roads, which seem to have been designed by the planners to give drivers a choice of reasons for arriving late. Either the layout has been dramatically altered, leaving a map of London as useful as a pork chop to a vegan, or each road has so many traffic-calming humps that the driver arrives feeling seasick and the wheels are splayed out at a thirty-degree angle. Why don't the road designers wake up all those sleeping policemen and put them back on the beat? That would stop road rage at a stroke.

Back to my trip, which, apart from the traffic, was superb. Stamford Hill Junior Mixed School, built in 1891, was still as it must have been then. It has certainly hardly changed in the last fifty-four years, except that it is now just a primary school. The classrooms and the hall are no different, and I'm certain I saw my old headmaster, Mr W.W. Ashton, sitting in his study. The first form teacher I had would be about seventy-five now, and that kind old Mr Baldwin, whose cricket whites lasted me for so many years, would perhaps be starting to fight his way through his eighties.

Tottenham County School took an age to reach through the friendly people of Seven Sisters Road and High Cross. (This is known as sarcasm. Most of the drivers here had a screw loose.) When we eventually got there it was to find that it is now the College of North East London. It appears to have swallowed up the fire station and swimming pool that were situated by the old school. I wonder whether anything, apart from Stamford Hill Junior, still exists. The salt beef bar where we had intended to have lunch no longer looks temptingly across at the Mecca betting shop I used to manage, which evens things up, as the betting shop is no longer there to be tempted. Along with those in Finsbury Park, Highbury Barn, Crouch End, Hornsey Rise, Grand Parade and so on, the betting shops all seem to have been bought by William Hill.

Undeterred and intent on finding some landmark well known to me that remained unchanged, we continued to eliminate the brick-built memories of my youth. In St Ann's Road number 90 had been Tottenham's answer to the Tower of Pisa. It was now an insignificant square of grass with not even a blue plaque to identify previous occupants, like the Gumbritches. The only part of that little area which was still there was the nursery school that bisected the site of number 90 and the corner of Plevna Crescent. The bottom half of Paignton Road, which had lived unobtrusively opposite 90 St Ann's Road for so many years, has been demolished and now sports a sign that says it is a park. I think it should require more than just a few square metres of brown grass to qualify it as a park. We walked back to the car along Holmdale Terrace, where Alan, a friend of mine until I moved to Harlow, used to live. Alan would be fifty-nine, the same age as me.

We drove to Highbury for a swift look around the Arsenal shop where Ren, despite the fact that he's a Tottenham fan, bought me a 2003/2004 handbook.

The roads seemed to be getting even more congested, if that was possible, so as we had arranged to meet my brother Pat in Harlow at 4.30, we had to miss a few of the calls I had listed, one of which was lunch.

When we arrived at River Way we drove around looking for the row of five shops and the big grass-covered hill that were both marker points for the road to the river. Unfortunately they had been demolished. Nevertheless, we soon located the bridge and crossed to the towpath that led to Latton Lock. The whole area surrounding the lock had been changed; what had been flat farmland was now a series of small lakes and hills. There has been a large number of pits dug and the soil excavated had completely altered the landscape. The holes were filled with water and some had joined together, appearing to form a second river at the back of the cottage.

A sign had been erected at the bridge, informing the user

that this was Latton Lock, but you would never have known a cottage had once been there. The brambles, bushes, nettles and trees were almost impenetrable from the first gate to the second, which were the owner's boundaries. The towpath was clear, but there was a wall of prickles to the side. However, Ren's Toetectors soon bludgeoned a path through to where we guessed the stairs to the cottage might be. Within fifteen minutes we had uncovered a large concrete slab that sadly was cracked across the middle and anyway was too heavy to carry. There was some excitement as the step was cleaned off and for the first time in years the sun shone on it, highlighting the date 1886. Built five years before Stamford Hill School. We were unable to take it with us, so we covered it again before we left.

The incredible discovery that I mentioned earlier was made just as we left the lock. Pat, with an amazed look on his face, pointed to a high bush. Growing up the plant and almost covering it was a large hop bine, with several tendrils leading off it. We walked back along the towpath, and all the way to the big bridge the bushes, and even some small trees, were almost engulfed by a carpet of hops. The only way they could have got there was if, when we went hop-picking from here in 1959, we had unknowingly carried some seeds in our clothing or furniture, and they had germinated next to the Stort and over the next forty-four years spread along the towpath. We didn't walk any further along the river, but even if it stopped growing where we crossed, that would still be at least half a mile of concentrated hop bines growing around the house of a family that used to go hop-picking regularly. Certainly nobody in that area of Essex is involved in cultivating hops. Those by the river have a slightly pink hue, but if left to their own devices, most living things will evolve and change.

We selected a couple of bines with the intention of displaying them in the kitchen, but needless to say we forgot them and they now reside in Pat's house. Typical that they should end up there, with the one person among us who hated hop-picking.

Ren and I stayed the night at Pat and Anne's in Harlow, and left at 8.30 a.m. the following day. We were twenty miles away when we remembered those hop bines at Pat's, but as we were going to Kent we assumed we would get some there.

The only problem arising from the drive to Canterbury was having to cross the bridge at Dartmouth in the shadow of a juggernaut, as we both dislike driving at altitude and the less the driver can see of the river, the better the nerves hold out. We got to Petham with almost enough time to see all I wanted to by lunchtime. The cricket pitch was as it had been, except it had two trees on it and I was sure there had been only one. Still, another one could have grown since we last came here.

The quiz question I had asked everyone a few months ago was the name of the pub in Petham. I received two replies, The Petham Arms and The Nelson. Well, the correct answer is – there isn't one. Sadly for Anita and Eileen, who so much enjoyed the aroma of the village stores, we couldn't find that either. There were several signs of yuppies in the village, but few signs of everyday life.

Ren and I decided to walk a little, so we parked by the front of the field where the black huts had been, and set off. As I had expected, the huts were no longer there and the field had been separated into three parts, one of which contained a flock of sheep. Had they been strolling about the field in the sixties, I think one or two of them would have been scrumped and roasted on a low spit, while a mug of mint sauce was brushed over it. We walked about the hills that had looked down on the huts, and the huge branches that the swings had been on still seemed to have the burn marks on them that were caused by the ropes.

At the bottom of the hills there was a row of small cobnut trees. The only way of telling what they were, if you didn't recognise the plant, would be through the huge number of nutshells littering the ground. We came to the manor house and gardens, which appeared unchanged, as did the oast house, although this had been extended to take another five or six homes. A little further along the lane the row of houses that were once used by the farm workers had been renovated. Little ancient farm implements decorated the gardens, while each house had now been christened Dun Roaming, or some such. It appears that the hardy hop-picker has been replaced by the Hooray Henry. The area where the red huts had been situated was now a corn field, which also encompassed the meadow that the bull had chased me and Ted across in 1954. Exciting times, but they never induced me to become a matador.

I think Ren was starting to get a little worried about my sudden swings from continual chatting to a near comatose condition. I was merely trying to recall some of the things that had happened on this farm in my youth.

We drove into Canterbury for lunch and noticed a sign advertising a beer and hop festival. That would fit in perfectly. Lunch would take us up to 3 p.m., then off to Faversham and the hop festival. As we drove through the outskirts of Canterbury, we saw a pub, with a turning opposite. We pulled into the turning and found ourselves in Kent cricket ground. Ren took some photos of the famous tree on the pitch and we crossed to The Bat and Ball, where we had some lunch. Ren had the all-day breakfast, although it didn't take him long to eat it. I ordered a hot chili beef that my mouth stood just three spoonfuls of before I swallowed a pint of shandy in an attempt to quench the flames. After removing ten small red chili peppers, I tried another mouthful, but it was still hot enough to melt my teeth so I did the only thing possible, put it down for the dog.

It didn't take long to find Faversham, but the festival consisted of an everyday little market, where the only reference to hops was from the farmer, taking up the width of the road with his trailer. He was selling hops at the astonishing price of ten pounds a bine. Considering the number of bines on The Banjo field, if he were able to sell them at the market, he would probably nett thirty thousand pounds on that field alone.

On the way back to Bournemouth we stopped at Fleet services on the M3. Ren bought us each a kiddies' Happy Meal, which entitled us to a free model footballer. I would have preferred a player from St Hares of the Harlow Sunday league, but I had to make do with David Beckham. Finally, arriving home from that memory-tracing trip, I felt deflated, much as parents do when their child goes off to school for the first time. Something had come to an end. All those places that had disappeared from my life: the homes, like 90 St Ann's Road, Tottenham, and Latton Lock Cottage, Harlow; the schools, Tottenham County and Netteswell Comprehensive; the places I worked, like Soper's farm, Harlow, and the betting shops around Stamford Hill and Tottenham. Not one of these exist any more.

There is, however, still one abiding monument to childhood. Clapton Common remains to remind me that, of all the many memories I have stored away, none would exist if this one had not ended as it did. The Common is an oval pond about two miles from where we lived in Tottenham. At one end stood a large hut, where a very pleasant man sold snacks. I was about five and was walking past the Common with Pat and Tom on the way to Springfield Park. Being rather smaller than my brothers, I was lagging behind a little. I have no idea what happened, I think I bent to pick a stone up and tripped, tumbling head first into the water. I must have completed a somersault, and I felt myself slowly sinking to the bottom. As I gently nestled in the weeds I clearly recall looking up through the slightly rippling water at the stunningly attractive sky. I know I was making no attempt to get out of the pond, although it was quite shallow. Suddenly hands were thrust down to grasp my shirt front and haul me out. I coughed and spluttered as Pat and Tom both whacked me on the back, then I began shaking.

The man from the café ran over to see what had happened, and took the three of us into his hut. While I hung my clothes in front of the fire, he gave us all tea and crisps (the real ones, with the blue bag of salt in). Once the clothes were dry, we thanked the man for his help and made our way home.

Shortly after that episode I went to the pool at High Cross and learnt to swim. But I sometimes wonder how long I would have remained admiring the sky if Pat and Tom had not been there to pull me out. I think the answer might have been, too long.

Some of the events in this book may have been witnessed by several people, but not remembered by all of them. There has to be a reason to be able to recall something fifty years after it happened. Either it was funny, or sad, or entertaining in some way. The group of us who spent so much time together did not set out to make memories, but the enduring friendships that we shared, and the carefree life we led, made it impossible to imagine we would ever forget any of the things that happened during those golden days in Petham.

But all memories, even the funniest, can come wrapped in a sigh. Those fifty years, where did they go?

It might help if I opened a waxworks museum to display these people and places, then whenever my spirits dipped, I could go there and chat to a room full of reconstructed candles!

Tally up, all full baskets!